Start
Rural Science

Bernard Salt

Head of Rural Studies
King Edward VI School, Lichfield
Chief Examiner, Midland Examining Group

CASSELL

Cassell Publishers Limited
Artillery House
Artillery Row
London SW1P 1RT

First published 1988

ISBN 0 304 31347 5

Design by Joy Fitzsimmons

Line drawings by Ian Foulis & Associates

Typeset by Best-set Typesetter Ltd

Printed in Hong Kong by Wing King Tong Co Ltd

Cover photograph of goats reproduced with the permission
of the Royal Agricultural Society

Photographs on pages 126, 128, 140 and 141 reproduced by
courtesy of the Forestry Commission

Contents

Preface

Start Rural Science is written for pupils in the lower part of the secondary school who take rural science either as a separate subject or as part of a general science course.

The units are designed to allow children to learn basic scientific methods and concepts whilst working with interesting materials to which they easily relate. Practical work is an essential part of this subject as it increases motivation and allows our pupils to have tangible and positive achievement. It is, however, difficult to undertake practical work with wild animals and here I recommend the use of some of the excellent audio-visual material currently available.

All the material in this book can be used with complete confidence as it has been tested in the teaching situation. The order in which the units appear has little significance; the sequence of work will be dictated by the seasons and a limited amount of cross-referencing is included to allow for this. It is hoped that pupils using this book will have an area of success together with an enjoyable learning experience.

Bernard Salt
Lichfield 1988

Unit 1
Classification of living things

Classification means 'putting things into groups or sets'. All living things can be fitted into five sets. These are:

| virus | bacteria | fungus | plant | animal |

This unit deals with the last two sets – plants and animals.

Task 1.1

Write down all the ways you can think of in which plants and animals are different.

The most important difference between plants and animals is the way in which they get their energy:
Plants get their energy from sunlight.
Animals get their energy from food.

The Plant Kingdom

We can easily divide plants into two sets:

Set 1: Plants which do not have flowers, e.g. seaweed (algae), moss and fern.
Set 2: Plants which do have flowers, e.g. grass, dandelion and oak.

Horsetails – an example of non-flowering plants

Above: Class 2 – dicotyledons

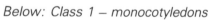

Below: Class 1 – monocotyledons

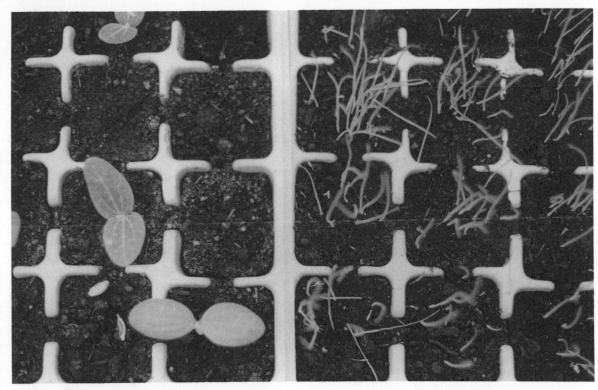

Young monocotyledons (right) and dicotyledons (left) growing from seed

The plants in set 2 (flowering plants) are easily divided into two further sets:

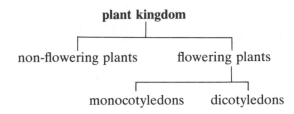

These two sets are called *classes*.

The plants in class 1 are *monocotyledons* (mono = 'one'; cotyledon = seed leaf). The plants in class 2 are *dicotyledons* (di = 'two').

Task 1.2

Look carefully at the three photographs above and opposite. Then copy the chart into your book and complete it.

The differences between monocotyledons and dicotyledons

	Monocotyledon	Dicotyledon
Food store in the seed	One	Two
Leaf veins – branched or parallel		
Seed leaf – one or two		
Flower parts	In 3s, 6s, etc.	In 4s, 5s or multiples

The Animal Kingdom

We can easily divide animals into two sets:

Set 1: Animals without a backbone – these are called *invertebrates*.

Set 2: Animals with a backbone – these are called *vertebrates*.

Stoat

Rabbit

Frog

Hen

Shrew

Wood louse

Pigeon

Butterfly

Spider

Fox

Snake

Task 1.3

Place the animals shown in the pictures into two sets – invertebrates and vertebrates.

Vertebrates are divided into five sets. These are also called *classes*.

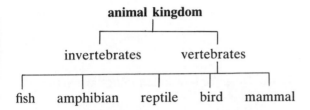

Characteristics of the five classes of vertebrates

Body covering	Breathing	Body temperature	Reproduction
Scales	Obtains oxygen from water throughout life	Varies with the temperature of the surroundings	Lays eggs and leaves them
Naked skin	Obtains oxygen from water during first stages of life only		Lays eggs and keeps them warm
Feathers and scales		Kept constant	
Hair	Obtains oxygen from air throughout life		Produces milk to feed young

Task 1.4

Copy this chart into your book. Write the correct class(es) of vertebrate in each empty box. (*Note*: some boxes will have only one class. Others will have more than one.)

Task 1.5

Classify the following list of vertebrates into five classes:
Stoat, adder, pigeon, tortoise, frog, shrew, sparrow, carp, lizard, newt, rabbit, toad, cow, slow worm, pike, chough, mink, grass snake.

Summary

1. All living things can be classified into sets and sub-sets. Sub-sets are given names such as *class* and *family*.
2. There are two classes of flowering plants – *monocotyledons* and *dicotyledons*.
3. There are five classes of vertebrates: *fish*, *amphibian*, *reptile*, *bird*, *mammal*.

Questions

1. Write single sentences to answer the following questions:
   (a) From where do plants obtain their energy?
   (b) From where do animals obtain their energy?
   (c) Name three plants which do not produce flowers.
   (d) In what way are reptiles different from amphibians?
   (e) How is it possible to tell which class a plant belongs to when only a leaf is available for examination?

2. (a) Describe how living things are arranged into sets.
   (b) Explain how knowledge of plant and animal classification can be helpful.

3. Horsetails, rose, oak, algae, moss, grass, buttercup
   (a) Arrange the above plants into two sets.
   (b) Add two more plants to each of your sets.

4. (a) Write down the names of all the vertebrate animals you can think of which live in this country.
   (b) Re-arrange your list into five different classes.

Unit 2
Vegetative structures

stem: the main support of the plant; it transports water and mineral salts from the roots to the leaves. The stem also transports the sugars, which are made in the leaves, from the leaves to the other parts of the plant.

node: the place on the stem where the petiole is connected.

leaf: usually green and flat. It 'breathes' air and uses it, together with the water and mineral salts from the roots, to produce the materials for plant growth. The energy used for this production of materials comes from the sun in the form of light.

roots: hold the plant in the soil. The root absorbs *water* and *mineral salts* (dissolved chemical substances) from the soil and transports them to the stem.

soil level

The external features of a plant

flower: the sexual part of the plant which will develop into a fruit, containing seeds.

internode: the part of the stem between one node and the next.

petiole: the stalk that joins the leaf to the stem.

bud: there is always a bud at a node; it is a compact cluster of tiny leaves on a very small stem. A bud will do one of three things: (1) produce a flower; or (2) produce a shoot; or (3) remain dormant throughout the life of the plant. (A dormant part of a plant is a part that is alive, but is not growing or changing in any way.)

The picture shows the external features of a plant. The parts which form flowers and seeds are called *reproductive*. The *seeds* grow into new plants. They each have their own food store so that they can survive from one season to the next (e.g. over the winter).

The parts of a plant which do not form flowers and seeds are called *vegetative*. Some vegetative parts can also produce a new plant. Many of these parts also have their own food stores. We will look at these in the next section.

Horizontal stems

Stolons
Stolons are horizontal stems which grow above the ground. They usually have long internodes. The buds at the nodes form roots and leaves and develop into new plants.

Rhizomes
Rhizomes are horizontal stems which grow underneath the ground. Some rhizomes (e.g.

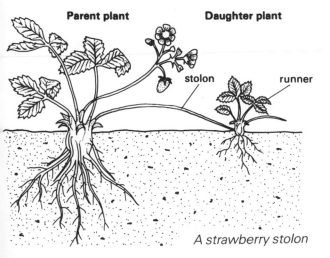

Parent plant **Daughter plant**

stolon runner

A strawberry stolon

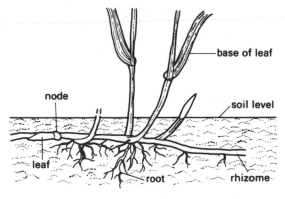

base of leaf

node

soil level

leaf

root rhizome

Couch grass rhizomes

iris) become swollen and fat with a food store. Others (e.g. couch grass) do not swell but remain slender and stem like.

Practical 2.1 Growing mint plants from rhizomes

(*September or October is a good time for this*)

Items required: mint plant; 9 cm plant pot; compost; knife

1. Dig up some mint. Examine the roots. Not all are fibrous; some grow horizontally and have buds on them:
   These structures are not roots – they are underground stems called *rhizomes*.
2. Cut five 4 cm lengths of rhizome, making sure that each one has a node.
3. Fill the pot with a soilless compost to within 4 cm of the top. Firm the compost gently.
4. Arrange the five rhizome sections evenly on top of the compost.

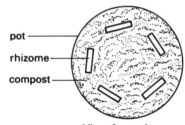

View from above.
before covering rhizomes with compost

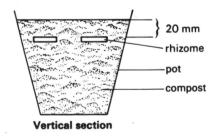

Vertical section

5. Fill the pot to within 1 cm of the top.
6. Water and place in the classroom window or in the greenhouse.
7. A few weeks later young mint shoots will be growing. If kept indoors they will provide mint during the winter.

Note: If you plant the mint in the garden it may become a weed which is difficult to control. Several of the most difficult weeds have rhizomes, couch grass for example.

Tubers

Some underground stems become swollen at the tip and form a *stem tuber*. The best-known stem tuber is the potato.

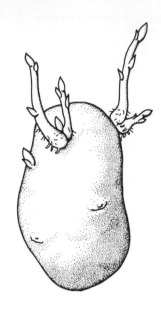

Roots also swell to form *root tubers*, for example the dahlia.

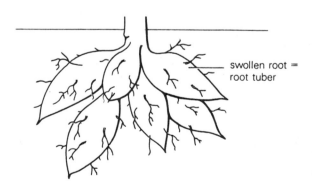

Root tubers do not have buds on them. They will not grow into new plants unless a small section of stem is attached.

Above: Stem tuber – the swollen stem of kohlrabi

Right: Root tuber – the swollen root of beetroot

Corms

Some plants produce swellings on short lengths of their vertical stem; these are called *corms*. On the outside, corms look like bulbs. Inside, however, they are solid flesh and do not have the layers seen in the onion. Some examples of plants which have corms are anemone, gladiolus and crocus.

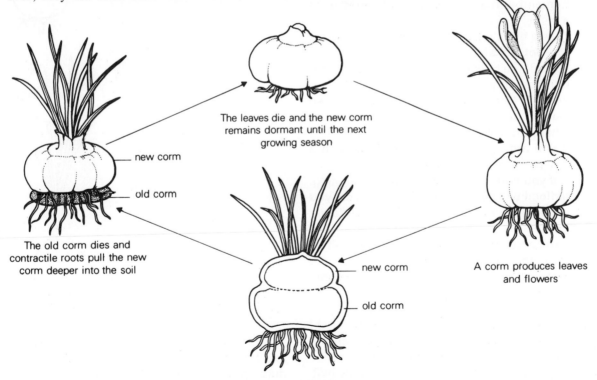

new corm

old corm

The leaves die and the new corm remains dormant until the next growing season

The old corm dies and contractile roots pull the new corm deeper into the soil

new corm

old corm

The flower dies and the leaves live on producing food; this food is stored in a new corm which grows on top of the old corm

A corm produces leaves and flowers

The life cycle of a corm

Practical 2.2 Investigating the life cycle of a corm

(*September or October*)

1. Dig up a dead gladiolus.
2. Cut through the stem about 4 cm above the corm. Remove the leaves.
3. Make a vertical cut all the way through the middle of the stem.
4. Refer to the diagram of the corm life cycle. Try to identify the parts on your specimen.
5. Draw the life cycle of a gladiolus corm. Start with a single new corm as it would be purchased from a shop.

Bulbs

Some plants store food at the bottom of their leaves. The swollen leaf-bases are called *bulbs*.

Practical 2.3 The structure of a bulb

Items required: an onion set

1. Remember that leaves and roots grow from a stem.
2. Take a bulb, look at it and observe the roots (or root remains).
3. Peel off the outside layer. This is a leaf; what purpose do you think it serves?
4. Peel off the next layers. These are also leaves, but quite different from the first; what purpose do you think they serve?
5. If you are careful you will finish up with a tiny piece of tissue – the stem.

   If you had used a bought flower bulb (e.g. daffodil or tulip) instead of an onion set you would have found an embryo (baby) flower inside.
6. Take a second onion set. Cut it in half to show the vertical section.
7. Make a large diagram of the vertical section. Label the parts.

Bulbs are produced by the plant as organs of *perennation* – that is, organs which will ensure that the plant survives the winter.

In spring when the weather is warmer the bulb will grow and produce flowers, which in turn produce seeds.

In the warm autumn soil, roots are produced.

In the depth of winter the bulb remains dormant.

In the warmth of early spring leaves grow.

In warmer April, flowers develop.

The life cycle of a bulb

Bulbs purchased in the autumn are planted and left throughout the winter in order to produce spring flowers. Before a spring bulb will flower it has to have two conditions: (i) a cold winter; (ii) a warm spring.

Bulb growers prepare some of their bulbs by keeping them in a refrigerator for several weeks in August and September. This is like an artificial winter and if these bulbs are brought into a warm house in the middle of winter they will start to flower.

Practical 2.4 Producing chicons from tap roots

(*first part done in June and July; second part done October/November*)

Items required: Witloof Chicory seeds; garden rake, line and spade; 22 cm pot; peat; black polythene or large cardboard box

1. Rake an area of dug soil. Sow Witloof Chicory seeds thinly in rows 25 cm apart.
2. Hand weed as necessary but make sure that the bed is weed free at the end of the summer term.

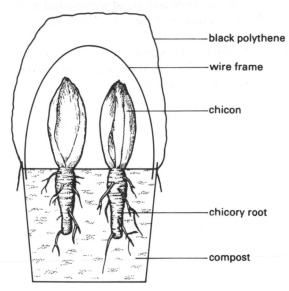

Growing chicons from chicory roots

3. In September the chicory will look very much like dandelion plants.
4. In October dig up ten plants from one end of a row. Take care not to chop the roots.
5. Examine the roots – these are '*tap roots*'; they are thick, long and vertical. Each root contains a food store.

6. Cut off the leaves as low as possible. Plant the roots in peat. Space them evenly with the tops just protruding above the peat.
7. Water and allow to drain.
8. Get a box or an *opaque* black polythene bag (take care: some black polythene is thin and allows light through). Cover the pot and tape it so that all light is excluded.
9. Keep in a cool place. Examine the growths by feeling through the polythene from time to time.
10. When the growths are about 12 cm high, take off the bag or box and examine them. Each root will have grown a 'chicon'. These are good to eat. They make a substitute for lettuce in winter. The other chicory tap roots can be lifted and grown as required.

Note: The box or polythene excluded light. This meant that the plant was unable to make its normal green colour and remained white. Celery, leeks and other vegetables have light excluded from certain parts to keep them white. This process is known as *blanching*.

Chicory

Summary

1. The parts of plants that are *not* flowers or fruit are called *vegetative*.
2. *Stolons* are horizontal stems which lie upon the soil surface.
3. *Rhizomes* are underground horizontal stems.
4. *Stem tubers* are parts of stems swollen with a food store.
5. *Root tubers* are parts of roots swollen with a food store.
6. *Corms* are swellings on short lengths of vertical stem.
7. *Bulbs* consist of a cluster of leaves on a short stem; some of these are swollen as food stores, while others are dead and protective.
8. A *tap root* has several purposes, one of which is to store food.
9. *Blanching* is the artificial whitening of a part of a plant by excluding light.

Questions

1. Describe the function (purpose) of each of the following: (a) root; (b) stem; (c) leaf

2. Use words and diagrams to explain why chopping off couch grass leaves with a hoe is not an effective method of control.

3. (a) Draw a large diagram of a potato and add the following labels:

   bud, scale leaf, eye, rose end, heel end

   (b) In what ways do stem tubers differ from root tubers?

4. Two hyacinth bulbs of the same variety appear to be identical. They are planted in pots during September and treated in a similar way. One hyacinth flowers at Christmas, the other flowers in March.

   Give a possible and *detailed* explanation of why the flowering times are different.

5. With words and diagrams describe how it is possible to produce a fresh supply of chicons throughout the winter.

Unit 3
Jiffy 7s

A Jiffy 7 compressed (right) and expanded (left)

A Jiffy 7 is a disc of compressed peat contained in a durable plastic net (see illustration, right). It has a diameter of 4.5 cm and is 1 cm thick. When soaked in water it expands to a height of 6 cm (illustration, left). It is widely used in horticulture for seeds, cuttings (see Unit 4) and seedlings.

Practical 3.1 Raising house plants from seeds using Jiffy 7s

Items required: Jiffy 7s; seeds; seed tray

1. Soak Jiffy 7s in a beaker, a few at a time, until they are fully expanded.
2. Pack 40 into a standard seed tray (five rows of eight) with the holes uppermost.
3. Sow one seed into the top of each Jiffy 7. Try to sow each seed twice the depth of its diameter. Cover with a little peat.
   (*Note:* When using very small seeds sow 3 or 4 and when the seedlings are large enough pull out all but the best one)
4. Keep in a warm classroom until the first seeds germinate.
5. Transfer to a warm greenhouse or place under a mercury vapour lamp.
6. Water, using a can with a fine rose, to keep the Jiffy 7s moist.
7. When the leaves of the seedlings begin to touch, pot up as follows:
   (a) Fill an 8 cm pot with potting compost until it is 7 cm from the top.

(b) Firm gently and add a little more compost if required.
(c) Place the Jiffy 7 complete with the growing plant into the centre (do *not* remove the plastic net).
(d) Fill up the pot with compost. Firm gently until the compost is 1 cm from the top.
(e) Grow on for a few more weeks watering as necessary.

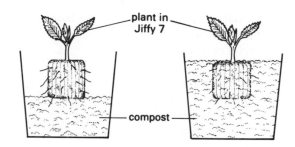

By this time you should have a good house plant. There are two ways of killing a house plant with water:

1. Not giving enough water.
2. Giving too much water.

The next experiment tests the idea that there is a relationship between water temperature and the time it takes a Jiffy 7 to expand.

Experiment 3.1

Items required: beakers; Jiffy 7s; stop clock; ice; thermometer

1. Add some ice to 500 ml of cold water and stir until the temperature is 0 °C.
2. Heat 500 ml of water to 70 °C.
3. Place eight beakers in a row. By mixing your hot and cold water fill them with water having a range of temperatures approximately equal to those in the diagram overleaf.

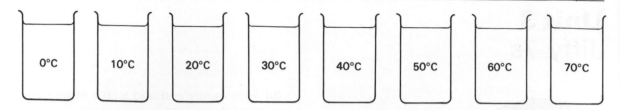

4. Start the clock and quickly place one Jiffy 7 in each beaker.
5. As the Jiffy 7s reach full size remove them from the water. Note the time each has taken to expand.
6. Plot your results on a temperature/time graph.
7. Join the points with a *line*.
8. Compare your graph with those below. Decide whether or not the idea has been proved correct.

Graphs like these would suggest that there is *no* relationship between temperature and expansion time:

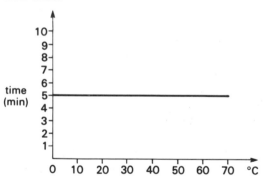

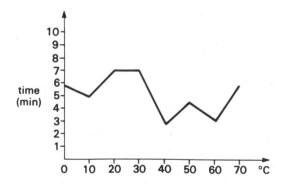

Graphs like this would suggest that there *is* a relationship between temperature and expansion time:

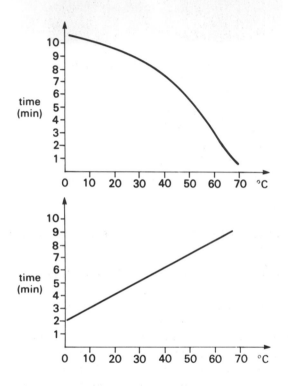

Summary

1. A *Jiffy 7* is a compressed block of peat in a plastic net.
2. Jiffy 7s are used to grow seedlings and cuttings.
3. Normally only one plant is raised in each Jiffy 7.
4. When potting up a Jiffy 7 plant the plastic net is not removed.
5. Jiffy 7s expand more quickly in warm water than in cold water.
6. Jiffy 7s break up in hot water.

Questions

1. Describe what a Jiffy 7 is and how you would use it to *either*
   (a) raise plants from cuttings; *or*
   (b) raise plants from seeds.

2. Jiffy 7s were soaked in water of various temperatures. The times taken for each Jiffy 7 to fully expand were as follows:

Temp (°C)	0	5	10	20	25	30	40	50	60	70
Time (min)	31	26	15	9	7	5	2	1	1	X

Note: at 70 °C the Jiffy 7s broke up.
(a) Use these results to draw a *line* graph.
(b) Explain the graph in words, using only a single sentence.

Unit 4
Cuttings

Some parts of plants when cut off can be made to grow into new plants. The new plants will have only one parent and will be a clone. A *clone* is exactly the same type as its parent.

Practical 4.1 Rooting shoots in water
(*best done in spring or summer*)

Items required: test-tube rack; large test-tubes; sharp knife

1. Almost fill the test-tubes with water. Cover the tops with aluminium foil.
2. Using a sharp knife, cut just below a node. Collect young vegetative shoots, each about 7 cm long, of one or more of the following: tomato (take a side shoot, not the top of the plant); coleus; willow; elder; tradescantia; ivy; impatiens (Busy Lizzie)

Conifer cuttings: these are taken in the autumn, kept over winter in a cold frame and will root in the spring

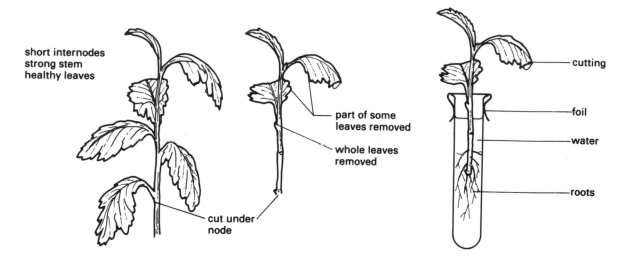

short internodes
strong stem
healthy leaves

part of some leaves removed

whole leaves removed

cut under node

cutting

foil

water

roots

3. Remove the leaves from the bottom half of the stem. Leave the leaves on the top 3 cm of stem.
4. Using a pencil, puncture the centre of the foil on each test-tube.
5. Put one shoot into each test-tube, so that half of its length is in the water.

6. Put the tubes on to a light window sill (but not in direct sunlight).
7. Examine them 2–3 weeks later. Are there any roots? If so, can you see any root hairs?
8. If these rooted cuttings are potted up into a compost they will have to produce a new set of roots. Roots which grow in water are different to roots which grow in soil.

Cuttings are not usually rooted in water. They root better and pot up more successfully if they have plenty of air around the cut end.

Plant growth is controlled by a chemical substance called a *hormone*. These chemicals occur naturally in the plant. The amount present at the base of a cutting can be increased by dipping it into rooting powder. This will make it root more quickly. Rooting powder consists of man-made plant hormones. It can be purchased from most garden centres. (*Note*: When using rooting powder tip a little powder on to the bench and dip the root into that; do not dip cuttings into the whole container of rooting powder as the damp will spoil it.)

Rooting Mediums

The substance in which a cutting is rooted is called a *rooting medium*. There are several different kinds including: silver sand, peat, vermiculite and perlite. One of the best rooting mediums is a mixture which contains 50% Cambark and 50% sphagnum moss peat. Jiffy 7s can also be used (see Unit 3).

The next experiment tests whether cuttings are more likely to root in perlite than in Jiffy 7s.

Experiment 4.1

Items required: plastic cups; perlite; Jiffy 7s; cling film; rooting powder; sharp knife; plants for cuttings

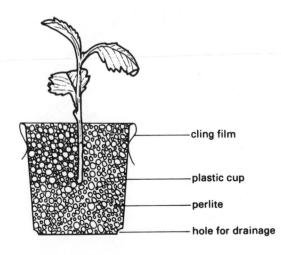

- cling film
- plastic cup
- perlite
- hole for drainage

1. Make drainage holes in the plastic cups by trimming off two 1.5 cm lengths from the bottom edge.
2. Soak the Jiffy 7s.
3. Fill the plastic cups with soaked perlite and cover the tops with cling film.
4. Using a pencil, make a hole through the cling film in the centre of each cup.
5. Prepare one cutting for each cup and one for each Jiffy 7. (*Note:* if you are using more than one species of cutting, half of each species should be in perlite and half in Jiffy 7s.)
6. Dip the bottom tip of each cutting into rooting powder. Insert the cuttings in the cup.
7. Put the cuttings into a warm greenhouse. Shade them with a sheet of green plastic, which stops about 40% of the light. They may also be left on the classroom window sill, providing it does not get any direct sun.
8. Leave them until some cuttings have rooted. (*Note:* To test for rooting, gently attempt to ease the cutting from the medium. If they have rooted you can feel the resistance offered by the new roots.)
9. Enter the results in a chart like this:

Number rooted in perlite	Number rooted in Jiffy 7s	Percentage in perlite	Percentage in Jiffy 7s

Having rooted your cuttings you may want to grow them on. Pot them up using a peat compost. See page 17 for instructions on potting up a cutting which was rooted in a Jiffy 7. To pot up a cutting rooted in perlite, break away the cling film and then lift out the cutting. Leave as much perlite clinging to the roots as possible and pot up in the normal way.

Other types of cuttings

The most usual kind of cutting is the growing shoot which you used in the last experiment. Some plants have other parts which will root and grow into new plants.

Sansevieria (Mother in Law's Tongue) and *Streptocarpus* (Cape Primrose)

These two common houseplants can be grown from a section of a leaf.

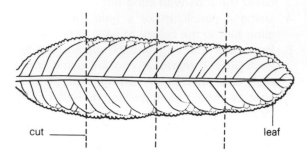

1. Take a whole leaf and cut it into 3 cm lengths.
2. Dip the bottom of each cutting into rooting powder.
3. Insert into rooting medium to a depth of 1 cm.

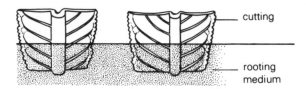

4. Keep warm and fairly dry until the cuttings have rooted.
5. Pot up and grow on in the normal way.

Note: If your *Sansevieria* has a yellow stripe down each edge of the leaf, the new plant which grows will not have this colouring; it will be all green. This is very unusual as plants grown from cuttings usually have the same characteristics as the parent plant.

A leaf cutting of Streptocarpus

just beginning to root

two weeks later

Saintpaulia (African Violet) and *Peperomia* (Rugby Football Plant)

With these plants the whole leaf is taken. Cut through the petiole, near to the bottom, with a very sharp knife. Dip the cut end into rooting powder. Carefully insert into the medium leaving the leaf (a).

Clamp the open end of the polythene bag between two rulers. Seal with a lighted match (under the supervision of your teacher) (b).

Stretch a line across a north-facing window. Peg the bag to the line with two clothes pegs.

Each leaf will produce roots and develop into a new plant which can be potted up.

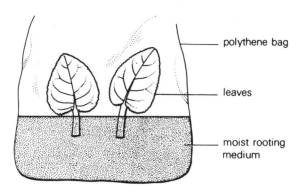

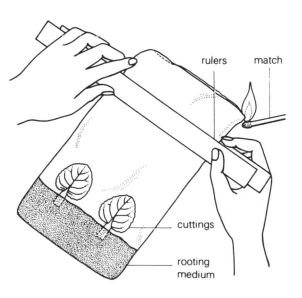

Yucca

These plants are on sale in nearly every large shopping centre. Have a good look at one (or examine the photograph). They are grown from cuttings which are just lengths of wood. Roots grow at the bottom and leaves at the top.
Note: If the wood had been inserted into the medium the other way up both leaves and roots would have grown from the bottom.

The needs of a cutting

If the roots or stem base of a cutting turn black it is because they have not got enough oxygen. Correct this by using a more airy medium and giving *less water*.

An exception to the above is the geranium. The geranium cutting requires different conditions from most other cuttings. After taking geranium cuttings they should be left in the sunshine for a few hours for a scar to form over the cut end. This treatment would kill most cuttings. The best time to take geranium cuttings is during the summer holidays. This plant is therefore not very suitable for school use.

Yucca

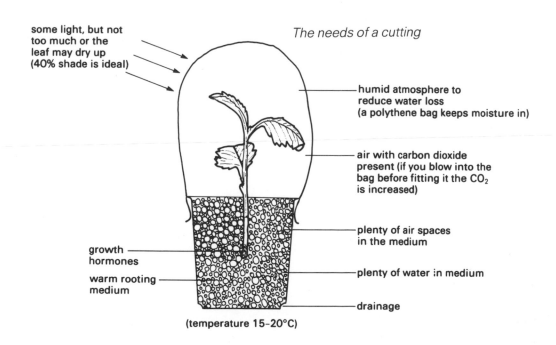

The needs of a cutting

some light, but not too much or the leaf may dry up (40% shade is ideal)

humid atmosphere to reduce water loss (a polythene bag keeps moisture in)

air with carbon dioxide present (if you blow into the bag before fitting it the CO_2 is increased)

plenty of air spaces in the medium

growth hormones

plenty of water in medium

warm rooting medium

drainage

(temperature 15–20°C)

Summary

1. Given the right conditions, some parts of plants may develop into new plants.
2. Cuttings require *light*, but not direct sunlight.
3. Rooting powder contains plant *hormones*. It makes cuttings root more quickly.
4. Cuttings require a *warm rooting medium* with lots of *air* and excellent *drainage*.
5. Cuttings require a *humid* atmosphere.
6. The area of leaf on a stem cutting should be small, but some leaves must be present.
7. Cuttings should be taken with a sharp, sterile knife.
8. Vegetative shoots are the most usual form of cutting, but some plants can be rooted from leaves and other parts.

Questions

1. Give a step-by-step explanation with words and diagrams of how to:
   (a) take cuttings from a named house plant;
   (b) root the cuttings;
   (c) pot up the rooted cuttings.

2. (a) List the conditions that a cutting needs if it is to root successfully.
   (b) Alongside each condition explain why it aids rooting.

   (c) Cuttings sometimes go black and rot at the base.
     (i) What is the cause?
     (ii) What could have been done to prevent this?

3. (a) Design an experiment to test a new rooting powder.
   (b) Describe the experiment in full. Include at least one diagram.

Unit 5
The fruit fly

The fruit fly (*Drosophila melanogaster*) is a very important insect for scientists. They learned about plant and animal breeding from studying it. They used the new things they had learnt from it to develop new cereal crops which have saved many parts of the world from famine. The fruit fly is also being used to benefit humans in many other ways, including the control of pests and diseases.

The next experiment tests the idea that wild fruit flies in this country would prefer to eat native fruits rather than imported fruits.

Experiment 5.1
(best done in September or October)

Items required: plastic cups; large glass beakers; cling film; a selection of native and imported fruits: e.g. damsons, apples, pears as examples of native fruits, and bananas, peaches, oranges as examples of imported fruits (over-ripe fruit is best as it gives quicker results, and may be obtained cheaply from a friendly greengrocer)

(The following instructions assume that the class has a range of fruits and each pupil is using one type only.)

1. Take enough fruit to half fill a plastic cup and mash it up thoroughly.

flies

beaker

plastic cup with mashed fruit

2. Put the mashed fruit into a plastic cup. Leave it exposed in a room with an open window for about 2 weeks.
3. When small flies can be seen over the cups, put each cup into a large beaker.
4. Two weeks later, trap the flies in the beakers by lowering a large plastic sheet over the jars. Taking care not to release the flies, cover your jar with cling film.
5. Your teacher will place a little ether-soaked cotton wool into your jar (or you may 'knock out' the flies by using a 'sparklets' soda syphon without water; the carbon dioxide from the bulb quickly kills the flies).
6. When your flies stop flying, tip them on to a piece of white paper and count them.
7. Record your results on a class chart like this:

Fruit	Number of flies per jar	Average
Apple		
Orange		
Banana		
Pear		

Fruit flies are one of the many living things which cause fruit to decay. By eating the flesh of the fruit the fruit fly allows the seeds to drop on to the soil where they may (or may not) grow.

Ripe fruit often goes mouldy. The mould is another living thing – a fungus – which is feeding on the flesh of the fruit.

Practical 5.1

Take one of your 'knocked out' flies. Examine it under the microscope. It may recover during your examination and fly away.

Note: Insects have three parts to their bodies:

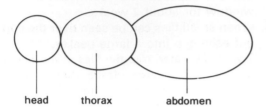

head   thorax   abdomen

Answer the following questions:

1. How many legs has it?
2. Which part of the body are the legs attached to?
3. What colour are the eyes?
4. What do the mouth parts look like?
5. Which parts of the insect have hairs?
6. How many wings are there?
7. Which parts of the body are the wings attached to?
8. How many segments form the abdomen?

Now use your answers to make a scientific diagram of a fruit fly. Your diagram will be better if you draw the outline faintly first and then add the details.

There are over one million different kinds of insects; that is about 80 per cent of all known animals. They are a very important part of nature and form part of most food chains. The vast majority of insects are neither helpful nor unhelpful to humans; a few, however, are – especially those which pollinate flowers as without them there would be little or no fruit. Also helpful are the insects which destroy pests. The unhelpful insects destroy about 20 per cent of the world's food either in the fields or in the store. Some insects spread disease to animals and to plants; millions of pounds are spent each year trying to control them.

Practical 5.2

Take your cup of decayed fruit and examine it. Answer the following questions:

1. Can you see any small cigar-shaped objects on the inside of the cup above the fruit? If so, take one and examine it under the microscope. What is it?
2. Look into the fruit. Can you see any small grubs? If there are any, examine one under the microscope. Are all the grubs the same size? What are they?
3. Take some surface material from the fruit. Examine it under the microscope. Can you find any eggs?

Task 5.1

Using the words: *adult, egg, larva* and *pupa*, draw a large diagram to show the life cycle of the fruit fly. Illustrate your life cycle with scientific diagrams.

Summary

1. The fruit fly is a wild animal that lives in this country.
2. The fruit fly larva feeds upon decaying fruit.
3. Insects have three body parts: *head, thorax* and *abdomen*.
4. Insects' legs and wings are attached to the thorax.

Questions

1. Describe an experiment which would test the idea that: fruit flies do not feed upon rotting vegetables; they feed only upon rotting fruit.

2. Draw a food chain diagram to show how the fruit fly transfers the energy contained in decaying fruit to an insect-eating bird.

Unit 6
The honey bee

The *honey bee* lives in a *colony*. Colonies may survive for many years. During the summer *pollen* and *honey* are stored in the nest; these stores enable several thousand bees to survive the winter.

In the colony there are three different kinds of bee:

The *worker* (below): numbers vary but 75 000 workers is not uncommon. The workers are sterile females; they live for only 6 weeks in the summer, but the ones born late in the year survive throughout the winter to work the following spring. The workers carry the sting.

The *drone* (above): several hundred males who live during the summer months only. In early autumn the other bees drive them from the hive. The drone has no sting. Drones are much broader than workers and make a different noise in flight.

The *queen* (above): a female with a life span of several years. The queen's sting is not barbed; the only time she uses it is when she kills other queens. The queen is easily picked out from the other bees by her long abdomen.

Wasp

Do not confuse the *wasp* with the honey bee. The common wasp is more slender than the honey bee and much more brightly coloured – it is a very distinctive yellow and black.

Task 6.1

1. List all the differences you can see in the four insects illustrated on this page.
2. Transfer the information from your list to a chart.

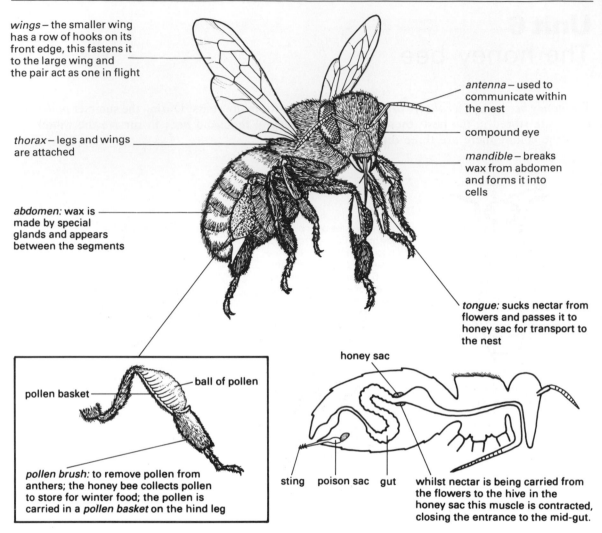

wings – the smaller wing has a row of hooks on its front edge, this fastens it to the large wing and the pair act as one in flight

thorax – legs and wings are attached

abdomen: wax is made by special glands and appears between the segments

antenna – used to communicate within the nest

compound eye

mandible – breaks wax from abdomen and forms it into cells

tongue: sucks nectar from flowers and passes it to honey sac for transport to the nest

pollen basket

ball of pollen

pollen brush: to remove pollen from anthers; the honey bee collects pollen to store for winter food; the pollen is carried in a *pollen basket* on the hind leg

honey sac

sting   poison sac   gut

whilst nectar is being carried from the flowers to the hive in the honey sac this muscle is contracted, closing the entrance to the mid-gut.

How a bee sucks nectar

Bees' nests

A bees' nest is built of hexagonal (six-sided) *wax cells*. These hang from the roof in vertical blocks of back-to-back cells. There is just enough room between the cells for the bees to work.

The cells have two uses:

1. They are used to store honey and pollen.
2. They are used to house the bees whilst they are developing from eggs to adults.

As there are three different-sized bees, there are also three different-sized cells.

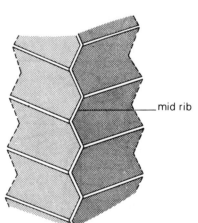

mid rib

Vertical section of comb
(note cells slope upwards)

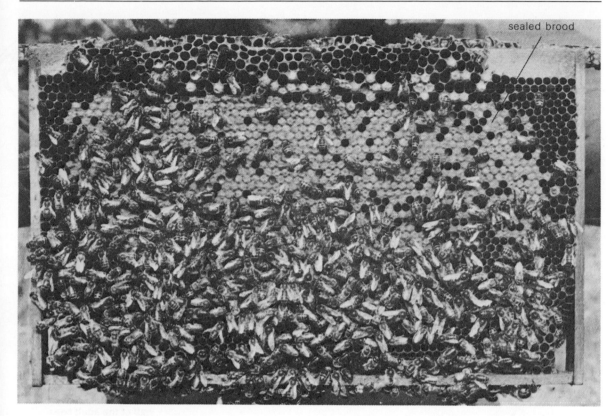

A bee comb

A vertical section of a comb (note: *the cells slope upwards*)

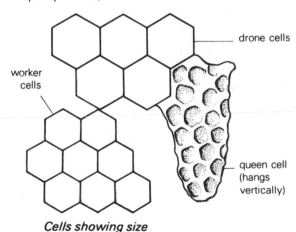

Cells showing size

Although bees are cold-blooded creatures they regulate the temperature inside their nests. Nests are warmed by workers eating and 'burning up' lots of honey. They are cooled by worker bees creating air currents. They do this by 'fanning' with their wings.

The honey bee is a social insect. Individuals do work for the benefit of the entire colony. Unlike most other insects, the larvae do not have to fend for themselves. They just remain in their cells and the worker bees keep them warm and fed. When they are ready to pupate, the workers cover the cell with a thin layer of wax.

Colony increase

Think of a bee colony not as thousands of animals, but as a single animal in which the queen and drones are the sex organs and the workers are the rest of the body. As more workers are produced the animal grows, but there is still only one animal. In order for reproduction to occur this single animal must have a baby. A colony's baby is a *swarm*; this swarm will eventually fly off and set up home on its own.

Larva development

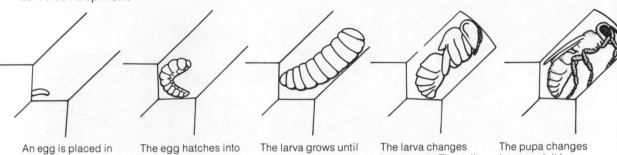

An egg is placed in the bottom of the cell by the queen.

The egg hatches into a larva and is fed by workers.

The larva grows until it almost fills the cell.

The larva changes into a pupa. The cell has been sealed with a cover of wax.

The pupa changes into an adult bee which breaks the seal and emerges from the cell.

Forming a new colony

One colony of bees

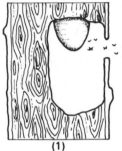

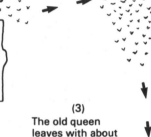

(1)
A colony of bees with one queen

(2)
A new queen is reared

(3)
The old queen leaves with about half of the adult bees in a swarm

(4)
The swarm flies to a new hole and begins to build comb for the queen to lay eggs in

Two colonies of bees

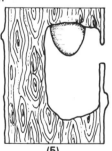

(5)
The new queen begins to lay eggs

A swarm

The bee sting

The *sting* is an adapted egg-laying tool. Therefore only females can sting.

The sting consists of three needle-like parts, two of which have barbs in their tips. As the bee stings the barbed portions are moved forward one at a time. This pulls the sting deeper into the victim whilst poison is being pumped in from the *poison sac*. Bees can easily withdraw their sting from other insects, but not from human skin. As the bee pulls itself away from a human victim its sting and poison sac are left behind.

A bee sting

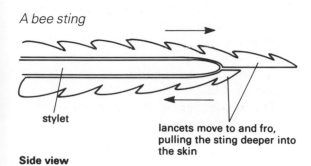

stylet

lancets move to and fro, pulling the sting deeper into the skin

Side view

stylet (pierces)

tube through which poison flows

barbed lancets

Cross section

Practical 6.1
(*best done in bright sunshine in June/July*)

Note: The ideal plant for this practical work is borage. A dozen plants sown in peat blocks or Jiffy 7s in January and planted out in April (60 cm apart) will produce a mass of flowers for many weeks.

Items required: stopwatch

1. Observe the insects on the borage plants.
2. Count the number of bee species. (In addition to the honey bee there are several different species of both humble bees and bumble bees.)
3. Select a honey bee and start the stopwatch. Count the number of flowers she visits. As soon as you lose sight of her stop the watch.
4. Record and repeat this for a number of bees. Enter your observations on a chart similar to this:

Bee no.	Time observed (seconds)	Number of flowers visited
Total	A	B

average time spent per flower $= \dfrac{A}{B}$

Sting in arm; note that the poison sac has been torn from the bee

If the sting is pulled out, more poison is squeezed into the arm

Remove the sting with a knife

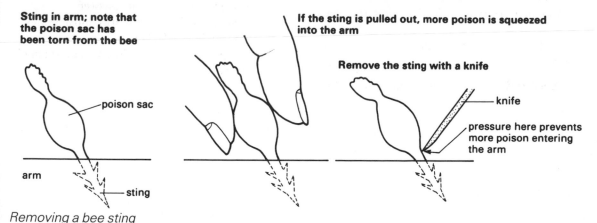

poison sac

knife

pressure here prevents more poison entering the arm

arm

sting

Removing a bee sting

You may have noticed that your bees visited only borage flowers. When bees are visiting flowers they visit only one species. This is also true of bees which are collecting pollen. You can confirm this for yourself. Stand alongside a hive and watch the insects returning with loaded *pollen baskets*. Each basket contains pollen of one colour only. This shows that the pollen came from only one species of flower.

Without pollination there would be no fruit nor any seeds. The bee is responsible for the largest proportion of insect pollination. Some pollination is done by moths, butterflies and other insects. But there is no other insect that keeps moving from flower to flower in the way the bee does.

Summary

1. Honey bees live in *colonies*.
2. There are three different types of bee in a colony: one *queen*, thousands of *workers* and (in summer) hundreds of *drones*.
3. Honey bees carry nectar to the colony in an internal sac.
4. Honey bees carry pollen to the colony in *pollen baskets* on their back legs. The pollen balls are held on by hairs.
5. Honey bees produce *wax* from which they make *cells*.
6. There are three different sizes of cells.
7. Cells are used for rearing young and storing winter food.
8. New colonies of bees are produced by *swarming*.
9. A bee cannot withdraw its sting from human skin.
10. The honey bee is the most important pollinating insect.

Questions

1. (a) Describe the three occupants of the bee colony.
   (b) What is the role of each of these?

2. Explain using words and diagrams how one colony of bees in spring may become two colonies by autumn.

3. A box which contained a colony of bees was weighed from time to time and the mass recorded:

   Recordings were as follows:

September 30th:	45 kg	March 4th:	25 kg
April 2nd:	25 kg	May 5th:	30 kg
June 1st:	50 kg	June 2nd:	45 kg
July 3rd:	55 kg	August 3rd:	50 kg

   (a) Explain:
      (i) the 20 kg fall from September to March;
      (ii) the 5 kg loss on 1–2 June.
   (b) Did any flowers bloom during March? Give reasons for your answer.
   (c) During which month was most honey and pollen collected?
   (d) Suggest a reason for the 5 kg fall during July.

Unit 7
Flowers

Rayless mayweed
(Matricaria discoidea)

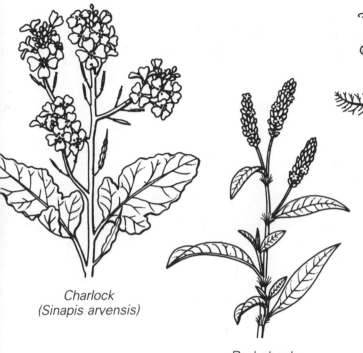

Charlock
(Sinapis arvensis)

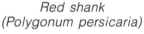

Red shank
(Polygonum persicaria)

Colt's-foot
(Tussilago farfara)

Scarlet pimpernel
(Anagallis arvensis)

Practical 7.1

1. Go to the garden and collect a number of flowers of different species. As in the diagrams here, they will appear to be very different from each other.
2. They all have petals and they all have a stalk. What other structures have they in common?
3. What flowers do have in common is purpose. The purpose of a flower is to give and/or receive pollen and then to form a fruit with seeds. Can you find a flower with a part which looks like a miniature fruit?
4. Examine the diagrams and captions on page 34. See if your flowers fit this general pattern.

The general pattern of a flower: structure and function

A flower consists of a stalk.

Vertical section   *View from above*

Around the stalk is a ring of 'leaves' called *sepals*. The sepals were originally the outside layer of the bud, and protected the developing parts of the flower before it opened. The complete ring of sepals is called a *calyx*.

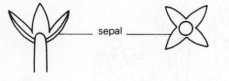

Vertical section   *View from above*

Further up the stalk is a second ring of 'leaves' called *petals*. The petals are usually the largest parts of the flower; they give it size, shape, colour and smell, attracting insects and giving them somewhere to settle. The petals often protect the organs inside the flower. The complete ring (whorl) of petals is called a *corolla*.

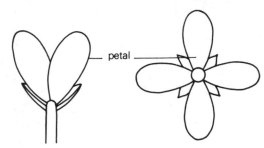

Vertical section   *View from above*

A little further up the stalk is a ring of stems, each stem holding a structure at its tip. These stems are called *stamens*. They are the male parts of the flower. Each stamen has two parts: the filament and the anther. The *filament* holds the anther in the best possible position in the flower; the *anther* produces hundreds of male sex cells called *pollen*.

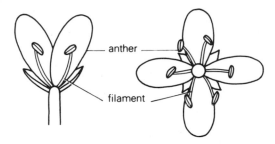

Vertical section   *View from above*

Finally, on top of the stalk is the female part of the flower. This has four separate parts: the stigma, style, ovary and ovules (which are inside the ovary). The *stigma* is where pollen from other flowers collects; it is often sticky so that the new pollen does not blow away. The *style* connects the stigma to the ovary. The style also holds the stigma in the best possible position to receive pollen from other flowers. The *ovary* is the container that holds the ovules. As the flower dies, the ovary begins to grow and will form a fruit, providing the flower has been pollinated (see page 36). An *ovule* is a tiny female sex cell. Each ovule will develop into a seed.

By selective breeding, some flowers with extra petals have been produced; these are known as *doubles* – a term which is often used in seed catalogues.

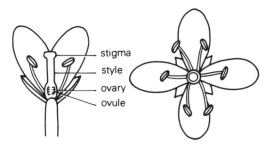

Vertical section   *View from above*

Some flowers have only male organs; others have only female organs. With this type of flower, only the female flower will develop into a fruit.

Male and female flowers

Some flowers have the ovary below the sepals.

Some flowers, like the sunflower, have a different structure to other flowers.

Practical 7.2

Items required: a sunflower flower or a dandelion flower

1. Take a sunflower or a dandelion. Have a good look at it.
2. If you are using a dandelion, take a single 'petal' and gently pull it from the flower.
   If you are using a sunflower, break the bloom in half and remove one of the many segments.
3. Examine the part you have removed. Tease it gently apart. You will see that it is in fact a single flower with sex organs inside and an ovary at the base.

The sunflower and the dandelion flowers consist of many individual flowers, all held together in what appears to be a large single flower. There are other flowers like this; they belong to one family called '*Compositae*'. The dictionary defines the word *composite* as 'a thing made up of various parts'. Can you see why this plant family is given this name?

Pollination

Pollination is the transfer of pollen from the anthers of a flower to the stigma of a flower. The flowers which are giving and receiving pollen must belong to the same *species*.

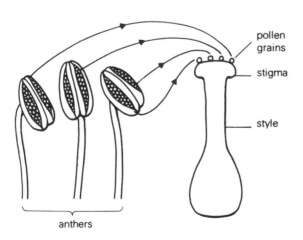

pollen grains

stigma

style

anthers

Self-pollination

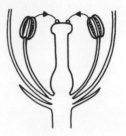

Self-pollination is the transfer of pollen from the anthers to the stigma of the same flower.

Cross-pollination

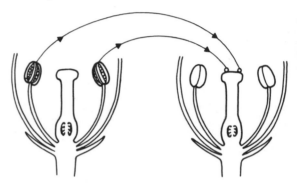

Cross-pollination is the transfer of pollen from the anthers of one flower to the stigmas of another.

Many flowers will not self-pollinate. They will only set (form) fruit with pollen from a different flower.

Insect pollination

The flowers you collected were most probably pollinated by insects. Flowers which are pollinated by the wind are not as obvious as they do not have to attract insects.

Insect-pollinated flowers often have large colourful petals. They also make *nectar* (a sugar solution) to attract insects. The pollen grains are fairly large and often a little sticky. The stamens and stigma are positioned so that they will come in contact with the visiting insects (see top diagram opposite).

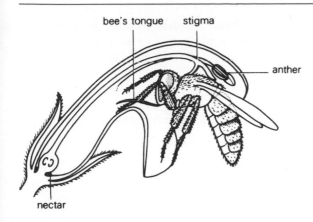

An insect-pollinated flower

Wind pollination

The pollen of some flowers is carried on currents of air. So the flowers do not have to attract insects. *Wind-pollinated flowers* do not produce nectar; their petals are small and not easily seen (e.g. grasses – see page 63). They produce large amounts of very light pollen from anthers which hang outside the flowers (see diagram). The stigma is often large, feathery and also hangs outside the flower. It is the pollen from these flowers which causes high 'pollen counts' in spring and problems for asthma and hay fever sufferers.

Task 7.1

Rewrite the following list in two columns, one headed 'insect pollinated' and the other headed 'wind pollinated'.

(a) The flower has a distinct scent.
(b) The flower has no scent.
(c) The flower has tiny green petals.
(d) The flower has large, coloured petals.
(e) The flower produces nectar.
(f) The flower does not produce nectar.
(g) Pollen grains are very small and light.
(h) Pollen grains are large and tend to stick together.
(i) Pollen is produced in fairly small quantities.
(j) Pollen is produced in very large quantities.
(k) Stigmas are small and hang inside the flowers.
(l) Stigmas are large and hang outside the flowers.
(m) Anthers are large and shake if touched.
(n) Anthers are small and firmly attached.

The next experiment tests the idea that courgettes will form only from flowers which have been pollinated.

Experiment 7.1
(*summer term*)

Items required: a number of courgette plants
  Sow the seeds in individual pots in March. Grow on, potting up to 22 cm pots.
  This experiment can then be performed in the classroom.

1. Take the courgette plants into the classroom. Cover the largest *female* flower buds

Note: The anthers and stigmas often ripen at different times – this prevents self-pollination

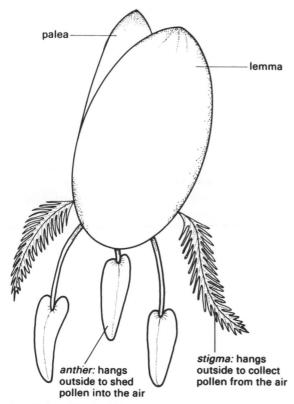

anther: hangs outside to shed pollen into the air

stigma: hangs outside to collect pollen from the air

A wind-pollinated flower

with muslin bags to prevent pollination by insects.
2. Keep them on the classroom window sill until the flowers open.
3. Uncover half of the bagged flowers. Gently stroke the stigmas with a *clean* brush or tail. Reseal the flowers and label the bags.
4. Remove each bag in turn from the other flowers. Using a rabbit's tail or a very soft artist's brush, transfer pollen from the anthers of a male flower to the stigmas of these flowers. Reseal the flowers and label the bags.
5. Examine the flowers 1 week later. Pay particular attention to the ovaries. Which are growing into courgettes? Which are beginning to wither?

Summary

1. A flower must be *pollinated* or seeds will not be produced.
2. *Single* flowers have one ring of petals. *Double* flowers have several rings of petals.
3. Some flowers have only male organs, and some have only female organs, but most have both male and female organs.
4. Flowers of the *Compositae* family are grouped together to form one large 'flower'.
5. The structure and pollen of *wind-pollinated flowers* are very different from the structure and pollen of *insect-pollinated flowers*.
6. Flowers will not form fruit unless they have been pollinated.

Questions

1. (a) Describe the differences between a wind-pollinated flower and an insect-pollinated flower.
   (b) List the reasons why pollen from insect-pollinated flowers has little effect upon the pollen count.

2. (a) What is the purpose of a flower?
   (b) Explain: (i) self-pollination; (ii) cross-pollination.
   (c) What effect(s) (if any) will a hive of bees, placed in an apple orchard at blossom time, have upon the apple crop in autumn?

3. Maize (sweet corn) is a wind-pollinated plant with male flowers on the top of its single stem and female flowers on the lower half of the stem. An amateur gardener found that he had high yields when he planted 25 maize plants in a block of five rows with five plants in each row.
   When he planted a single row of 25 maize plants he had low yields.
   (a) Draw a diagram to show the two different planting schemes.
   (b) Write a letter to the gardener explaining why his yields are better when he plants his maize in a 5 × 5 block.

Unit 8
The sweet pea

The plant upon which the sweet pea grows is similar to the garden pea. They both belong to the *legume* family. The flowers of the sweet pea are larger, more colourful and smell sweeter than those of the garden pea. Sweet pea seeds are *not* good to eat.

The seed of a dicotyledon

All seeds of dicotyledons have a similar structure. The sizes and shapes differ but the basic structure does not.

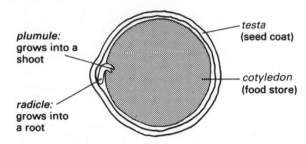

Dicotyledon seed structure

A seed has three main parts:

1. A tiny plant with a root, which is called a *radicle*, and a shoot, which is called a *plumule*.
2. A food store, in two pieces, called *cotyledons*.
3. A tough protective skin, which is called a *testa*. The testa has a breath pore, which is called a *micropyle*.

Germination

The next experiment tests the idea that sweet pea seeds germinate more quickly if the seed-coat is damaged.

Experiment 8.1
(*best done in November or December*)

Items required: sweet pea seeds; 7.5 cm plant pots; a peat compost; sandpaper

See diagrams on page 40
1. (a) Take two pots and fill loosely with compost.

   (b) Level the top.

   (c) Tap on table until compost is 3 cm from the top.

2. Take 10 seeds and carefully rub each one on sandpaper. Do this until a few square millimetres of the outside skin has been removed. (Do *not* sandpaper the area near to the small black scar.)
3. Evenly space the seeds on the top of the compost in one pot. Space 10 seeds which have not been sandpapered in the other pot.
4. Label both pots with your name, the date and whether it contains treated or untreated seeds. A piece of paper taped on the side of the pot makes a good label.
5. Fill both pots level to the brim with compost. Carefully press it down until it is 1 cm from the top.
6. Water well and leave in the warm classroom.

When the seeds germinate (begin to grow) record the numbers in each pot on a chart like this:

Date	Number germinated (untreated seeds)	Number germinated (sandpapered seeds)

Record other pupils' results on your chart as well as your own. The more results you have the better. Scientific experiments should be repeated many times before any conclusions are reached. Record your conclusions from this experiment in your laboratory note book.

As soon as the seeds have germinated put them into a cold greenhouse or a garden frame. (A *cold* greenhouse is unheated.) Check them each week and water as necessary.

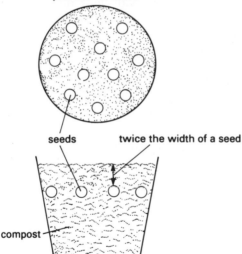

Top view showing position of seeds

seeds

twice the width of a seed

compost

Section through the pot

PHILIP ELKIN
SANDED

PHILIP ELKIN
UNSANDED

Watering plants requires great care: too little water and they die, too much water and they die also. In winter plants need very much less water than they do in summer and extra care needs to be taken. Do not water your sweet peas until they are almost dry and then water by filling the pot to the rim once only. It is possible to tell when the pots are dry by one of the following ways:

1. The plant is wilting (if this is the case you are late; plants should never be allowed to wilt).
2. The surface of the compost appears very dry.
3. The pot feels light when lifted from the bench.
4. If the plants are in a clay pot – tap it with a piece of wood. A high note means the plant needs water, a low note means that it does not need water.

FOOD FOR LARGE POLLINATORS

SWEETPEAS

SOIL IMPROVEMENT BY BACTERIA
WHICH LIVE IN ROOT NODULES

Growing sweet peas is good for the soil and for wildlife

When the sweet peas have two or three pairs of leaves they will need planting into separate pots. This is known as *potting-up*.

Practical 8.1 Potting up

Items required: 20 7 cm square vacuum-formed plastic pots; 2 carrying trays; peat compost
Note: If larger pots are used the plant will not make enough roots to keep the compost loose and 'open'. There is a danger that the pots will become water logged and the plants will then die

1. Put 2 cm compost into each pot.
2. Carefully tap out the ball of compost with the growing seedlings.
3. Carefully crush the compost ball and separate the seedlings. Leave as much compost clinging to the roots as possible.
4. Pick up a seedling with one of its leaves, hold it in the centre of a 7 cm pot and carefully fill up with compost.
5. Press the compost firmly around the edges of the pot. Add more compost if required. Leave 1 cm of the pot unfilled; this space is needed when the plant is watered.

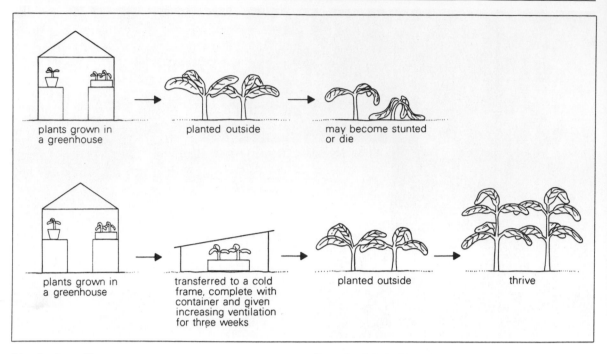

plants grown in a greenhouse → *planted outside* → *may become stunted or die*

plants grown in a greenhouse → *transferred to a cold frame, complete with container and given increasing ventilation for three weeks* → *planted outside* → *thrive*

Hardening off

6. Pot up the other seedlings and return them to the greenhouse.
7. Water the seedlings and check them regularly.
8. Towards the end of March put the seedlings outside, with some protection from north and east winds, to 'harden off' (get used to outside conditions). They will not be harmed by snow or frost as they are very *hardy* plants.

A *hardy* plant is one which will withstand outside conditions.

A *half-hardy* plant is one which will withstand outside conditions during summer when there are no night frosts.

Practical 8.2 Planting out

Items required: 5 1.5 m garden canes; string; trowel

1. Prepare an area of soil (taking great care not to step on the soil in which plants are to be grown). Mark out a circle with a diameter of 1 metre.
2. Stick in the garden canes around the outside of the circle. Adjust them until they are equal distances apart.
3. Push each cane hard into the soil (they will blow over if you do not). The finished job will look better if all canes are pushed in to an equal depth.
4. Draw the tops together and tie firmly.
5. Using a trowel, dig five holes on the inside of the circle near to the canes. The holes should be just large enough to take a 7 cm pot.
6. Carefully tap out each plant. Take great care not to break the ball of compost or damage the plant.
7. Place each plant in a hole keeping the surface of the ball level with the soil surface. Fill the space around the ball with soil and firm gently.

at your sweet peas and you will see that the following parts are green: stems, leaves, petioles (leaf stalks), buds, flower stalks and tendrils (the thin growths which are holding the plants on to the string).

Except for the petals, every part of the plant is collecting light for growth. Note how the leaves are thin and flat; this shape is a good one for catching light. Tendrils also collect light for growth. They hold the plant up high, which means that it is less likely to be overshadowed by tall weeds.

The next experiment tests the idea that a sweet pea plant responds to having its flowers cut off by producing more flowers.

Experiment 8.2

1. Do not cut the flowers on one of the sweet pea wigwams.
2. Gather all the flowers from the other wigwams as soon as they are mature enough (when the flowers at the bottom of the stalk are showing colour and those at the top are still in bud).
3. After a few weeks compare the numbers of flowers on the harvested wigwam with those on the unharvested wigwam.

Observe and record what happens to the flowers on the unharvested wigwam. During the last week of term, compare the number of flowers fit for gathering on this wigwam with the numbers on the other wigwams.

Practical 8.3 Observing flower structure
(*see Unit 7 for pollination*)

Take a fully open single bloom and answer the following questions:

1. Is the flower an attractive colour?
2. Does the flower have a sweet smell?
3. Colour and smell attract insects to the flower; what *two* items may be collected from the flower by insects?
4. How many sepals has it? What is (or was) the purpose of these?
5. Remove the large standard petal and the two next largest petals.

8. Tie string around the canes as shown in the photograph.
9. Inspect every week or so. Remove any weeds by hand.
10. At first the sweet peas will not appear to be growing; this is because the plant is using most of its energy to grow roots.

    After a few weeks strong growths will appear. Carefully tie these to a cane or a string to start the plant growing upwards. Your plants will now climb without any further help.
11. Cut the flowers as soon as they show colour. Use scissors and cut at the bottom of each flower stalk to give the longest possible stems.

The green parts of a plant help it to grow. Plants only grow when they are in sunlight. Take a look

6. The part you have left is rather like the keel of a boat. Hold the stalk in one hand and press a pencil on top of the keel as shown below. What do you observe?

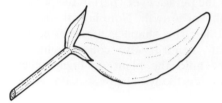

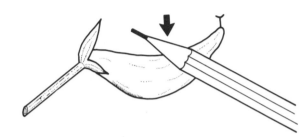

7. It would take a large insect to press the keel down. Which insects do you think are the most likely pollinators of sweet peas?
8. The *keel petal* is formed by two petals joined together. Carefully remove these two petals.
9. You are now holding the male and female flower parts. Remove the male parts (the stamens which make pollen). How many are there?
10. What you have left are the female parts: the stigma, style and ovary (see Unit 7 page 34).

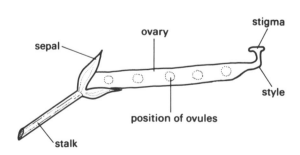

Practical 8.4

Take a second flower and dismember it. Using tape, stick the parts into your book. Label them and then describe the purpose of each part.

After the summer holidays there will be lots of sweet pea seeds. These can be harvested and used. The sweet peas from home-saved seeds will contain a high proportion of blue ones. If no new seed purchases are made for a few years the whole crop will probably be blue.

Practical 8.5

When you have finished with your sweet peas, carefully dig up one or two plants. Gently remove the soil by shaking or washing. Can you see any wart-like growths? Make a drawing of the root to show the growths. Label the growths as *nodules*.

Nodules contain bacteria which produce a natural fertiliser for the plant and help it to grow.

The legume family (peas, beans and some others) all have nodules on their roots. These contain bacteria which increase the nitrogen content of the soil. Growing legumes improves the soil.

Root nodules

Summary

1. When the green parts of plants are in sunlight they assist growth.
2. Sweet pea seeds may germinate more quickly if the seed-coat is damaged.
3. Cutting sweet pea flowers causes the plant to grow more flowers.
4. Peas and beans belong to the *legume* family.
5. Planting a seedling into an individual pot is known as *potting up*.
6. Sweet peas are pollinated by bees.
7. Sweet peas have tendrils which help them to climb.
8. A *hardy plant* is one which can withstand cold weather.
9. Getting a plant used to outside conditions is known as *hardening off*.
10. Legumes have *nodules* on their roots. These contain bacteria. Growing legumes improves the soil.

Questions

1. (a) Explain, using words and diagrams, how a sweet pea flower is pollinated.
   (b) The anthers of a sweet pea flower ripen at a different time to the stigma of the same flower. Is this flower more likely to be cross-pollinated or self-pollinated?

2. The sweet pea plant has a fairly soft stem. Give a detailed description of how this plant reaches heights of 3 metres.

3. Give a detailed description of how sweet pea plants are raised from seeds.

4. 100 g of sweet pea seeds were tumbled in sand in order to chip the seed-coats. Then they were sown in trays in a greenhouse. A second batch of 100 g were sown without being tumbled in sand.
   Results:

Date	Number of seeds germinated			
	Tumbled seeds	*Total*	*Untreated seed*	*Total*
11 Nov	8	8	0	0
18 Nov	459	467	26	26
25 Nov	464	931	386	412
2 Dec	69	1000	322	708
9 Dec	0	1000	380	1088
16 Dec	0	1000	12	1100

(a) If there are 12 seeds per gram, calculate the percentage germination for each group of seeds.
(b) Construct a bar graph to display the results.
(c) What conclusions can be drawn from this experiment?
(d) This experiment was repeated three more times. What is the point of repeating this experiment?

Unit 9
Cereals

Cereals are types of *grass*. We have bred them to produce large seeds. These are used for food for ourselves and our animals. They are *wind-pollinated monocotyledons*.

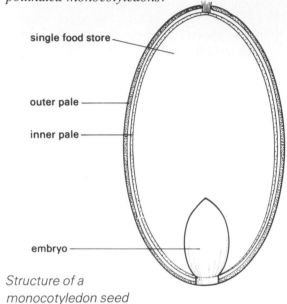

single food store

outer pale

inner pale

embryo

Structure of a monocotyledon seed

WHEAT – flour, bread and animal food

There are six main types of cereals: *wheat, maize, rye, oats, barley* and *rice*.

MAIZE – 'Corn on the cob', silage for cattle

BARLEY – for brewing beer and feeding animals

RYE – grown for grazing in early spring

OATS – porridge, cooking, and feeding animals

RICE – not grown in this country

Small quantities of cereal seeds can be difficult to purchase. Pet shops may have them as hen food, but they are often mixed.

Most agricultural seeds may be obtained in small quantities from: Cade Horticultural Products Ltd, Cade St, Heathfield, East Sussex TN21 9BS.

Practical 9.1 Germination testing

Items required: cereal seeds; paper towels; cling film; seed trays

1. Soak six paper towels.
2. Cover the bottom of a seed tray with the soaked towels.
3. Count exactly 100 seeds.
4. Place the seeds on the towels in neat rows.
5. Cover with cling film.

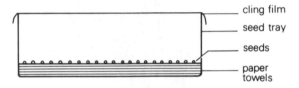

6. Leave in a warm room for a few days.
7. Count the number of shoots which are emerging.

This number is equal to the percentage germination of the seeds. NB. If the percentage germination is low then extra seeds will have to be sown to obtain the correct plant population.

Conditions for germination

In order to germinate all seeds must have the following conditions:

1. Plenty of *moisture.*
2. Plenty of *air* (seeds use oxygen as they germinate).
3. *Warmth* (normal room temperature is usually ideal).

Some species require other conditions as well as the three above. These include: light; passing through a bird's gut; a few weeks of very cold weather; a number of years' *dormancy*; there are even some seeds which will not germinate until they have been partially burned.

Fortunately cereals require only the first three conditions. The following experiments look at some of these.

First, let us look at the condition of warmth. The next experiment tests the idea that it is better to store seeds at a low temperature than at a high one.

Experiment 9.1

Repeat the practical above using two seed trays. In one tray use seeds that have been stored in the warmest possible place (a school classroom or an airing cupboard – *not* in an oven or you may kill the seeds).

In the other tray use seeds which have been stored in a refrigerator.
Note: The seeds should have been stored for at least 2 months.

Now we will investigate the condition of moisture. The following experiment tests the idea that it is better to store seeds in a low humidity than in a high one.

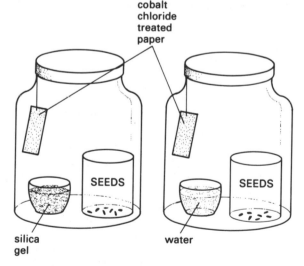

1. Take two screw-top jars and set them up as shown in the diagram. The water in one will create a humid atmosphere. The silica gel will absorb moisture from the air. The cobalt chloride is an indicator of air humidity: it is

blue when the air is dry and pink when it is moist.

The seeds in the two jars must be taken from the same sample.

2. Leave the jars side by side for at least 2 months.
3. Measure the germination percentage of both lots of seeds as in the practical above.

After both experiments are complete, answer the question: In what environmental conditions should seeds be stored?

Seed storage is very important as seeds form such an important part of our food and most seeds are only produced during a very short period each year. Also nearly all food crops are grown from seeds (the exception to this is potatoes, which are grown from root tubers; see Unit 3); these have to be stored from harvest until sowing time. In some parts of the world, poor storage of seeds is often the cause of famine. This is particularly true in hot climates where storage is more difficult.

Practical 9.2
(*best done in October*)

Items required: cereal seeds; rake; large plastic bottles with the bottoms cut away

1. Take about 2 g each of wheat, barley and oat seeds.
2. Rake an area of dug soil the width of a rake head and a metre long.
3. Mark three circles with the bottom of a bottle evenly across the raked area.

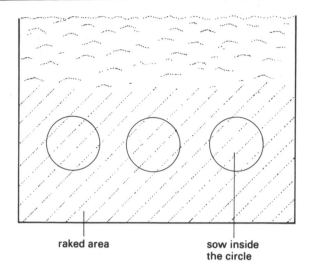

raked area   sow inside
                    the circle

4. Remove a handful of soil from one circle, sprinkle one type of seeds evenly and then cover with the handful of soil.
5. Press the bottle over the seeds firmly into the soil; remove the top.
6. Repeat for the other two types of cereals.
7. Three weeks or so later, when the cereals are about 10 cm high, remove the bottles.

Note: It is impossible to grow a small quantity of cereals without protecting them from the birds. Methods other than plastic bottles may, of course, be used.

Some farmers use a 'gas gun' to keep birds away. This is effective for a week or so until the birds become used to it.

A gas-gun bird scarer

Practical 9.3

(to be carried out when the growing cereals are in the fifth leaf (or more) stage of growth)

Items required: 10 cm by 15 cm piece of sticky-back plastic

1. Carefully remove one whole cereal plant from each of the growing circles and take them to the classroom.
2. Examine the plant. Has it got a tap-root system or a fibrous root system?
3. Follow the diagrams below and identify the leaf blade, ligule, auricle and sheath.

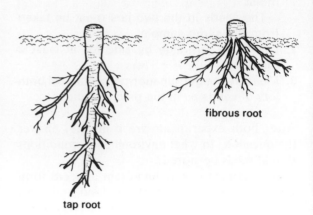

fibrous root

tap root

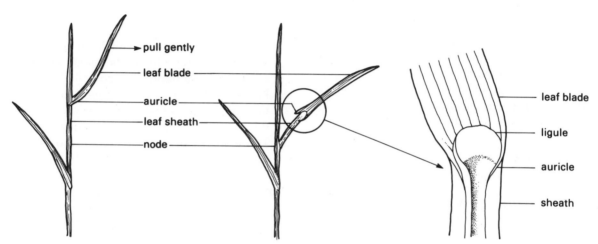

Wheat – auricles are hairy

Oats – auricles are very small

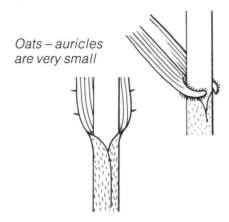

Barley – auricles have no hair (barley is bald)

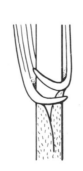

4. Can you find a bud at the node as in dicotyledons?
5. Using the diagrams identify each one of your plants.
6. Take one plant and a piece of sticky-back plastic which is slightly larger than the plant.
7. Carefully display the plant on a page of your book. Add as many labels as possible.
8. Remove the backing from the plastic. With great care stick it over the plant to make a permanent record.

The small differences in these three cereal crops may not seem important but small differences can make a big difference to the way in which a crop is used. For example oats are much better for feeding to horses than either wheat or barley. Oat straw is a good winter fodder for cattle, barley straw may be used but wheat straw is no good at all.

Yields are also different; average yields are:

wheat	5.2 tonnes per hectare
barley	4.1 tonnes per hectare
oats	4.0 tonnes per hectare

The world record wheat crop was grown in Scotland:

world record: 14.0 tonnes per hectare

During the first part of the summer term the cereal plants will come into flower. At first glance the flowers look similar to the ears of grain illustrated at the beginning of this unit.

Practical 9.4
(*best done when yellow anthers are hanging outside the cereal flowers*)

1. Remove a single 'grain' from an 'ear' of one of your cereals. This is not a grain of cereal: it is a flower.
2. Pull away the two outside protective layers – the glumes.

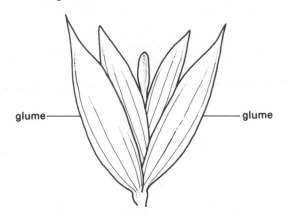

glume — glume

3. There are a number of structures inside. These are the individual flowers. Using two large pins, open a single flower.

Identify the parts from the diagram:

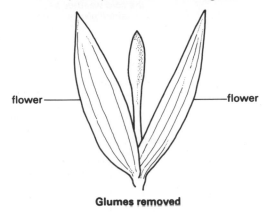

flower — flower

Glumes removed

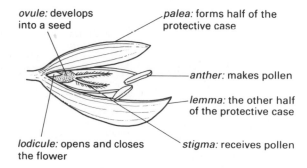

ovule: develops into a seed

palea: forms half of the protective case

anther: makes pollen

lemma: the other half of the protective case

lodicule: opens and closes the flower

stigma: receives pollen

During the summer term your cereals will grow ears. Unless you protect these from the birds you will soon lose the seeds as they form. The author's pupils regard this as a piece of conservation and enjoy watching the birds taking the grain.

Now that you have grown your own cereal you will probably have a good idea of the life cycle of a cereal. This is shown in the diagram overleaf.

Summary

1. *Cereals* are *grasses* with large seeds used for human and animal food.
2. There are six main types of cereals: *wheat, maize, rye, barley, rice* and *oats*.
3. Seeds are best stored at *low temperatures* and *low humidity*.
4. Cereals have a *fibrous* rooting system.
5. Cereal plants can be identified from their *auricles*.
6. Cereal flowers are *wind pollinated*.

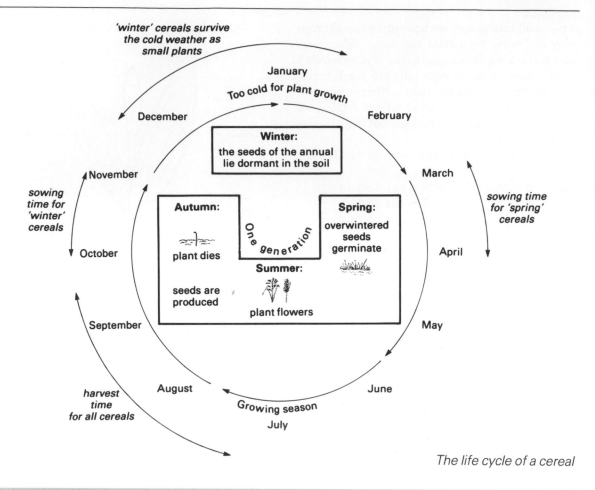

The life cycle of a cereal

Questions

1. Describe an experiment which would test the idea that it is better to store seeds in the dark than in the light.

2. Make a large copy of this chart. Enter as many uses in the empty boxes as possible. (A visit to the library will help you with this question.)

Cereals	Uses
Oats	
Wheat	
Rice	
Barley	
Maize	

3. Cereals are wind pollinated. Sometimes the anthers hang outside the flowers; sometimes the stigmas hang outside the flowers. But these organs never hang outside together.
   (a) List all the features of the cereal flower which shows that it is a wind-pollinated flower (refer to Unit 7).
   (b) Why do the anthers hang outside?
   (c) Why do the stigmas hang outside?
   (d) Why do the anthers not hang outside at the same time as the stigmas?

Unit 10
Growing cereals

Almost four million hectares of cereals are grown in this country each year. This is four times more than all other crops (except grass) added together. The main cereals are wheat and barley.

The reason for growing cereals is to harvest the seeds. These contain a lot of *carbohydrate*, which is an energy-giving food and a necessary part of our diet.

In some areas the *straw* (dead cereal stems and leaves) is important as bedding for live-stock, insulation for stored crops and fodder for cattle (NB wheat straw is unsuitable for fodder). In areas where few cattle are kept the straw is a nuisance as it chokes the plough and other machinery. To overcome this problem the straw is burnt in the fields.

One hectare (10 000 square metres) of wheat produces over 5 tonnes of grain. This is equivalent to half a kilo per square metre.

Straw burning

A tractor-drawn plough

Task 10.1

1. Suppose that you eat 250 g of wheat each day (as bread, biscuits, cakes, etc.). Calculate the area of land needed to grow enough wheat just for you.
2. Using a calculator, find the square root of the area in part 1. (This figure is equal to the side of the square which occupies that area of land.)
3. Go to the sports field and, using string and garden canes, mark out the area. Have a good look at it.
4. Measure the length and breadth of the sports field. Calculate its area. Divide this area by your answer to part 1. This gives the number of people that your sports field could grow wheat for.

Machines used for growing cereals

Many of the machines in the following sections are pulled by a *tractor* (see photograph below).

The plough

The purpose of the *plough* is to turn over a layer of surface soil. This buries the weeds and stubble. It also increases the amount of air in the soil.

This turns a single furrow and will plough a narrow strip.

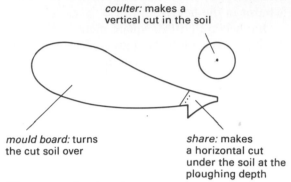

coulter: makes a vertical cut in the soil

mould board: turns the cut soil over

share: makes a horizontal cut under the soil at the ploughing depth

The parts of a plough

A modern plough will plough several furrows at the same time, and a wide strip at each pass. As the mould board is set it can only turn the furrow over in one direction (say left to right). The plough on page 53 has two sets of mould boards; one turns left to right, and one turns right to left. Whilst one set is ploughing the other set is in the air. At the end of the furrow:

1. The plough is raised.
2. The plough is rotated to bring the working mould boards to the top and the others to the bottom.

The harrow

The job of the *harrow* is to break up the ploughed soil until it is fine enough to give good conditions for germination. There are many different kinds. The one the farmer selects will depend upon the type and condition of the soil.

Some harrows are just pulled along.

Some harrows are powered by the tractor.

A harrow with moving parts powered by a tractor

A harrow that is simply drawn by a tractor

The seed drill

The *seed drill* loosens the soil with its front part, and then places rows of seeds and fertiliser in the soil. These flow down into the seed rows through sprouts leading from a box on top. A harrow drawn behind the drill covers them up.

Seeds need to be sown very accurately. The aim of most cereal growers is to grow 300 evenly spaced plants on each square metre.

A seed drill

The roller
The *roller* is used on some types of soil only. It improves the seedbed, controls some pests and pushes stones into the soil where they will not damage cutting machinery.

The sprayer

The *sprayer* adds liquid chemicals to the field. A cereal crop is sprayed a number of times during its growing period. The chemicals which are used include:

1. *Selective herbicides:* these kill the weeds but not the crop.
2. *Fungicides:* these kill fungi. There are many fungus diseases of cereals. If these diseases are not controlled the whole crop may be lost.
3. *Insecticides:* these kill insects. The most common insect pest is a sucking insect called the *aphid* (greenfly). The grower collects ten cereal ears and if there are more than 50 aphids on them in total he will probably spray. If there are less than 50 aphids he will probably leave the crop unsprayed.

A tractor-drawn sprayer

A cereal trailer being fitted at a grain store

A combine harvester

The combine harvester

This machine is called a *combine harvester* because it 'combines' several operations:

1. It cuts the crop.
2. It gathers up the parts that it has cut off.
3. It separates the grain from the rest of the plant.
4. It stores the grain in an internal tank.
5. It drops the plant remains (straw) on the ground.
6. It empties its tank into a trailer.

The trailer

The *trailer* carries grain from the field to the store, or to the drier.

A spreader

The spreader

Most cereal growers use a lot of fertilisers to help the crops grow. The *spreader* puts a thin layer of fertiliser evenly over the ground. The photograph shows half-tonne bags of fertiliser being loaded into it. This is usually done at the end of the season, after the crop has been harvested.

Once cereal grain is harvested it may need to be stored for some time. The next experiment investigates storage conditions. It tests the idea that the keeping quality of grain depends upon its moisture content.

Experiment 10.1

Items required: 10 plastic cups; 1 kg of wheat, barley or oats; cling film

1. Number ten plastic cups 1–10.
2. Weight 100 g of cereals into each cup.
3. Using a graduated syringe, add water to the cups as shown below. The cup with no water added is the control.
4. Cover each cup with cling film, place your hand over it and shake it carefully to disperse the water. Do *not* spill any.
5. Leave the cups in a warm classroom. Inspect them each week.
6. Record your observations.
7. State your conclusions.

 nil 2 ml 4 ml 6 ml 8 ml 10 ml 12 ml 14 ml 16 ml 18 ml

In a dry summer, grain can be stored without drying; in a wet summer, the grain may need to be dried or it will not keep in store. The most usual method of drying is to blow dry air through the grain. As the grain loses water it also loses weight. If 100 tonnes of grain has a water content of 13 per cent there is a total of 13 tonnes of water. If the farmer dries the grain until the water content is 12 per cent he has lost 1 tonne of water. His grain now weighs 99 tonnes.

This problem of *water content* will appear a number of times in your rural science course. It is overcome by using the idea of *dry matter* (DM).

A grain dryer

100 tonnes of grain with a 13 per cent water content can be illustrated as in the pie chart.

The grain is said to have a *dry matter content* of 87 per cent. How much water would 100 tonnes of grain with a DM of 85 per cent contain?

If grain were dried to the point that it contained no water at all it would be too hard to use. All grain contains some water – *usually* around 12 per cent.

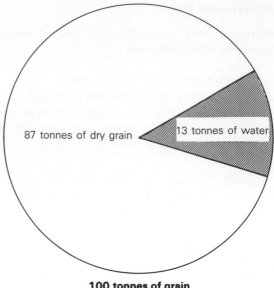

100 tonnes of grain

Summary

1. *Cereals* are the most important *arable* crop in the UK (arable = ploughed land).
2. Machinery is used in cereal production in the following order: *plough*; *harrow*; *drill*; *sprayer*; *combine harvester*; *trailer*; *spreader*.
2. *Selective herbicides* are chemicals which kill weeds without killing the crop.
3. *Fungicides* are chemicals which kill fungi.
4. *Insecticides* are chemicals which kill insects.
5. An *aphid* is an insect which sucks plant juices.
6. Moist grain (above 15%) will not keep in store.
7. The mass of grain minus the mass of water it contains is known as the *dry matter* content.

Questions

1. (a) **Using words and diagrams, describe the plough and say what its purpose is.**
   (b) **Why do modern ploughs have two sets of plough bodies?**

2. **Chemical sprays can be dangerous. Why then are they sometimes sprayed over food crops? (Give at least three reasons.)**

3. **50 tonnes of grain has a moisture content of 16 per cent.**

   (a) **What is the mass of dry matter?**
   (b) **If the grain has to be dried to 12 per cent before being put into store, what is the mass of water which will have to be removed?**

4. **Describe how a crop of wheat is grown. Explain the purpose of each machine you mention.**

Unit 11
The wood pigeon

This unit refers to the *wood pigeon* (*Columba palumbus*). The pigeons which live in large cities are not wood pigeons but are descended from the rock dove (*Columba livia*). This very common bird is lighter in colour than its city cousin. Its feathers are a mixture of blue/grey and shiny purple and green. There are two white patches either side of its neck; these are important as they help flying pigeons to recognise pigeons feeding on the ground.

Food

The pigeon's strong beak allows it to exploit a wide range of plant foods, including newly sown cereal seeds, ripening heads of cereals, cereal seeds shed in the stubble, weed seeds (fathen and charlock are preferred), charlock and cultivated mustard leaves, tree fruits (acorns and beech mast), flowers, leaves and buds of trees, clover, brassica leaves, peas at all stages of growth from seedling to ripe pod, and many other plants, both wild and cultivated.

Pigeons feed in flocks which often number over 1000 birds, and at night the birds roost in trees and hedgerows. Each bird requires either 225 g of fresh leaf or 45 g of seeds each day (or a mixture of these) to remain alive and healthy.

A brussels sprout plant damaged by pigeons; every plant in a 7 hectare field had been damaged in a similar way

This pigeon was shot as it flew to roost; the contents of its crop, alongside, have been removed, and consist entirely of brassica leaves

Task 11.1

A study of pigeon diet was made over a period of 1 year, in an area of Lancashire with the following results:

January	Brassicas	(20)	Clover	(60)	Weeds	(10)	Tree seeds	(10)
February	Clover	(50)	Brassicas	(30)	Weeds	(20)		
March	Cereals	(30)	Weeds	(30)	Clover	(20)	Brassicas	(20)
April	Cereals	(50)	Weeds	(20)	Tree flowers	(20)	Clover	(10)
May	Weeds	(70)	Clover	(20)	Legumes	(10)		
June	Weeds	(90)	Cereals	(5)	Legumes	(5)		
July	Cereals	(50)	Weeds	(30)	Legumes	(20)		
August	Cereals	(90)	Legumes	(10)				
September	Cereals	(90)	Legumes	(5)	Tree fruits	(5)		
October	Cereals	(85)	Tree fruits	(15)				
November	Cereals	(60)	Tree fruits	(40)				
December	Tree fruits	(60)	Cereals	(30)	Brassicas	(10)		

Percentage of total food intake is shown in brackets

Using graph paper, transfer the above information to a colour chart. Draw a key alongside to explain the colour code. An example of the representation of this information is shown below:

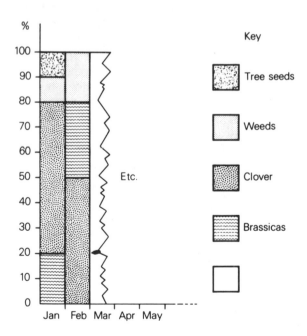

It takes the pigeon about 1 hour to collect 45 g of wheat and 8 hours to collect 225 g of clover leaves. So it follows that, when wheat is available, the bird will spend very much less time feeding than when clover is the food source. Pigeons feed more rapidly in the late afternoon as they need to fill the crop before flying back to roost. Birds shot as they return to roost have very full crops.

Breeding

Pigeons build rough nests of twigs in trees and hedgerows. Two eggs are laid. The female incubates them for 17 hours each day, and the male incubates for 7 hours whilst the female feeds. If the large white eggs are left uncovered they are quickly taken by predators. These include jays, magpies, jackdaws, rooks, squirrels, rats, stoats and weasels. During the 17-day incubation period, 50 to 60 per cent of pigeon's eggs are taken by predators. Losses of *squabs* (baby pigeons) are much less than egg losses, due partly to the aggressive display and hard peck given by the young.

Both male and female pigeons produce a milk-like substance on the lining of the crop. The squab pushes its beak down its parent's throat and this milk is pumped from parent to young. Pigeon milk is all that the young receive for the first 3 days of life. Afterwards parents carry normal food to the nest to supplement the milk. About 22 days after hatching, the young leave the nest to collect their own food, but they return daily for a further week to receive some food from their parents. Before the young leave the nest, the parents build another nest, lay a second clutch of eggs and begin incubating them.

Breeding continues from May to September; the largest number of nestlings are reared during August (78 per cent) when food is plentiful. On average one pair of pigeons will rear three young each season. Many of the young die as juveniles; those that survive the first year live to an average age of 3 years, but many live longer than this. There is a record of a wild pigeon that lived for 14 years.

A pigeon-scaring device

Control

The pigeon population in the UK remains fairly constant. It is naturally controlled by the amount of food available in winter. For every two birds in spring there are five by autumn, so if the population is to remain constant three of these five must die during the winter from starvation or other causes. Any birds that are shot during the winter were either doomed to die or leave food upon which others will survive. Unless more than three-fifths of the total pigeon population can be shot during the winter, shooting will not reduce the pigeon population. Although much pigeon shooting does take place each winter, numbers killed are less than those required to reduce the population. The population would be reduced (for a short time) if shooting took place at the beginning of the breeding season. But at this time of year the birds spend little time feeding and much time in the trees. Here they are hidden from view, making shooting impractical.

Experiments with drugged wheat seeds have been carried out. The seeds are spread in fields where pigeons feed, and any bird eating the seed is 'put to sleep' for an hour or so. During this hour pest species are collected and destroyed, the other birds (game birds and protected birds) are left to recover. At present no suitable drug has been found to make this a satisfactory method of control.

The only effective method of control is to cover the crop with netting to prevent wood pigeons from reaching it, or to scare off any pigeons that come to feed. The first method is only practical where very small numbers of plants are involved (e.g. a row of spring cabbages in a garden). The second method works well when there are plenty of other available food sources, but if food is scarce the flocks will return and bird scarers become less effective. Many different types of bird scarers are available, including gas guns (see Unit 9), bangers, stuffed hawks, balloons filled with light gas and models of men holding guns. All these devices are effective when first positioned. However, birds soon become accustomed to them. When hungry, they feed close by them. Bird scarers are most effective when a crop is susceptible to damage for a short period (e.g. newly sown cereals); by the time the birds cease to react to the scarer the crop is no longer vulnerable to attack.

The pigeon-scaring device in the photograph on page 61 is blown around by the wind. Alternate panels are painted fluorescent orange and black; this gives the appearance of flashing. The crop being protected is kale.

Summary

1. The pigeon is a herbivore with a varied diet.
2. Pigeons nest in trees and hedgerows. They lay two eggs in each brood.
3. A pair of pigeons will rear two or three broods each year.
4. Pigeon shooting is unlikely to reduce the pigeon population.
5. The pigeon is considered by cereal and vegetable growers to be a pest.
6. Bird-scaring devices are not effective against pigeons for more than a few days.

Questions

1. Write single sentences to answer the following questions:
   (a) How do flying pigeons recognise other pigeons which are feeding on the ground?
   (b) Where do pigeons go at night?
   (c) What is a 'squab'?
   (d) Why do bird scarers work for only a few days?
   (e) What do newly hatched pigeons eat?
   (f) Why are so many pigeons' eggs taken by predators?

2. Describe a year in the life of a pigeon. Include in your description the types of food which will be eaten during the year.

Unit 12
Grass

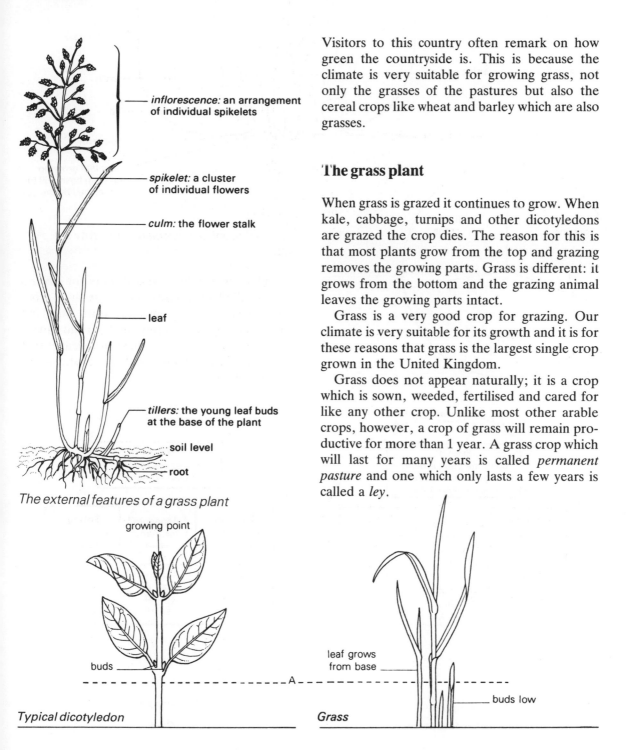

inflorescence: an arrangement of individual spikelets

spikelet: a cluster of individual flowers

culm: the flower stalk

leaf

tillers: the young leaf buds at the base of the plant

soil level

root

The external features of a grass plant

growing point

buds

Typical dicotyledon

A

leaf grows from base

buds low

Grass

Visitors to this country often remark on how green the countryside is. This is because the climate is very suitable for growing grass, not only the grasses of the pastures but also the cereal crops like wheat and barley which are also grasses.

The grass plant

When grass is grazed it continues to grow. When kale, cabbage, turnips and other dicotyledons are grazed the crop dies. The reason for this is that most plants grow from the top and grazing removes the growing parts. Grass is different: it grows from the bottom and the grazing animal leaves the growing parts intact.

Grass is a very good crop for grazing. Our climate is very suitable for its growth and it is for these reasons that grass is the largest single crop grown in the United Kingdom.

Grass does not appear naturally; it is a crop which is sown, weeded, fertilised and cared for like any other crop. Unlike most other arable crops, however, a crop of grass will remain productive for more than 1 year. A grass crop which will last for many years is called *permanent pasture* and one which only lasts a few years is called a *ley*.

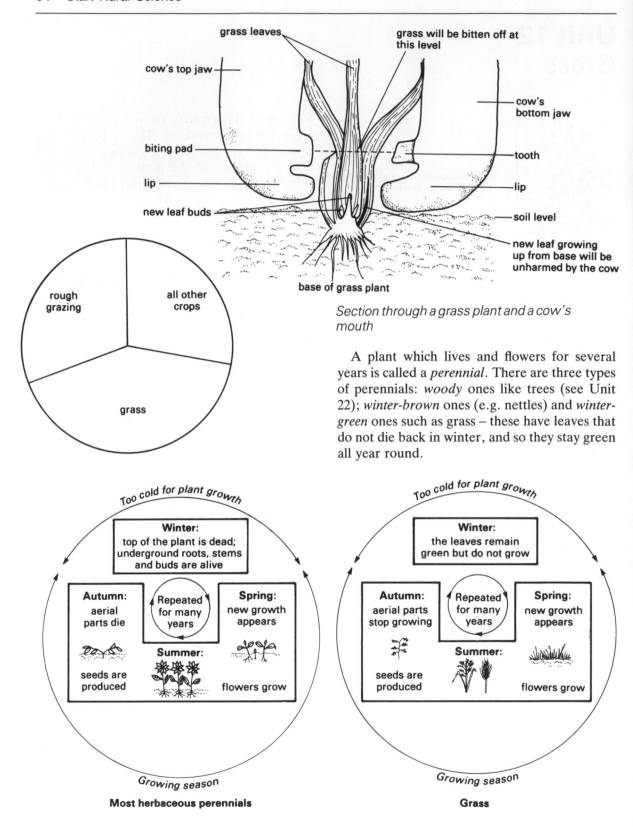

Section through a grass plant and a cow's mouth

A plant which lives and flowers for several years is called a *perennial*. There are three types of perennials: *woody* ones like trees (see Unit 22); *winter-brown* ones (e.g. nettles) and *winter-green* ones such as grass – these have leaves that do not die back in winter, and so they stay green all year round.

Most herbaceous perennials

Grass

Practical 12.1

Items required: two quadrats held apart with 7 cm wires soldered into each corner; sheet of perspex; chinagraph pencils

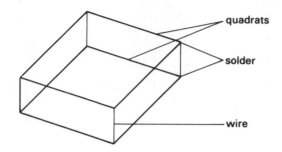

1. In a pasture field (or use the sports field) make a random throw with the double quadrat.
2. Examine the plants growing inside the quadrat. Are they all grasses?
3. Place the perspex on top of the quadrat. Using a different colour for each, shade over:
   (a) areas of grass;
   (b) areas of clover;
   (c) areas of other plants;
   (d) bare areas.
4. Find a clover plant (they are abundant in most pastures) and dig it up very carefully. Is it joined on to other clover plants? The horizontal stems are stolons (see Unit 2).
5. Take the perspex to the classroom. Estimate the percentages of grass, bare ground, etc. inside your quadrat.
6. Use your results to construct a bar chart.
7. Compare your chart with those from other areas.

Note: The clover in the sports field was probably not sown. White clover is an indigenous (native) plant which grows wild.

Practical 12.2 Estimating the proportions of grass, clover, weeds and bare ground in a pasture (or a playing field)

Items required: metre rule

1. Number 1–100 in your note book.
2. Take a metre rule.
3. Place the rule on the pasture and look at the 1 cm mark. If a grass plant is growing alongside put a 'G' by number 1 in your note book. If a weed is growing alongside put a 'W'. If it is bare ground put 'B'; if it is clover put 'C'.
4. Look at the 2 cm mark on the ruler. By number 2 in your note book, put 'G', 'W', 'B' or 'C', as in part 3.
5. Continue in the same way to the end of the rule.
6. When you have finished, count the number of Gs, Ws, Bs and Cs. These numbers are the percentages of each type of plant present in the pasture. If the average is taken from a whole class the results will represent a fair estimate for the field.
7. Display your results on a pie diagram. To find the angle of each, multiply the percentage by 3.6.

Most of the weeds you find are low growing and unlikely to be grazed. Some like thistle and nettle may be large. These survive because the cow will not eat them. Why not?

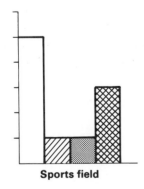

Sports field

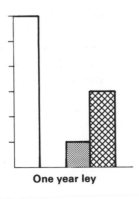

One year ley

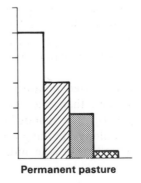

Permanent pasture

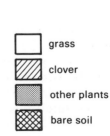

grass

clover

other plants

bare soil

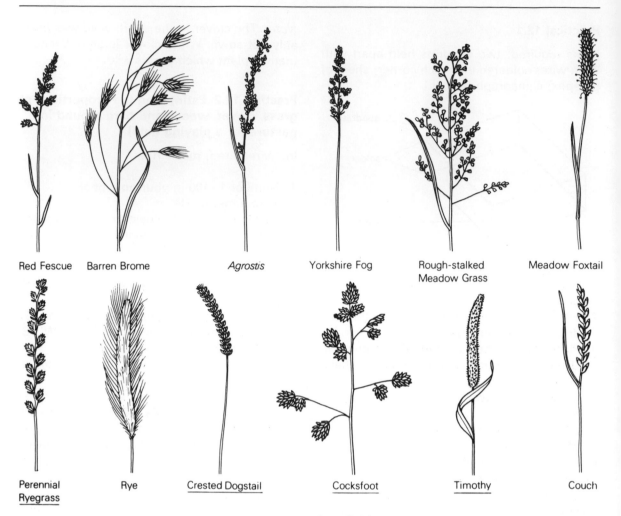

Red Fescue	Barren Brome	*Agrostis*	Yorkshire Fog	Rough-stalked Meadow Grass	Meadow Foxtail
Perennial Ryegrass	Rye	Crested Dogstail	Cocksfoot	Timothy	Couch

Examples of grasses; those underlined are sown in farm fields

Grass species

Grass is not a single species. There are over 10 000 different species of grass in the world; over 150 of these grow in the United Kingdom.

Only a few grass species are productive enough to be sown in fields for grazing hay or silage.

When a pasture is sown a 'seed mixture' is usually used; it includes more than one grass species and white clover is very often included in the mixture – this will grow alongside the grass. Clover is a legume (see Unit 8) and is included for two reasons:

1. Clover improves the soil and makes the grass grow faster.
2. Clover plants contain protein and minerals in greater amounts than grass plants. So eating some clover with their grass helps to balance the cows' diet.

The growth curve of grass

Examine the graph below and answer these questions:

1. During which months of the year is there a surplus of grass?
2. During which months of the year is there no growth?

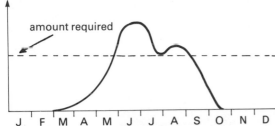

The actual times of growth and no growth will vary from year to year and from district to district. In all districts, however, there is a period of surplus and one of shortage. This is a problem for the farmer – his cattle will have too much food in some seasons and none at all in others. This problem is overcome by keeping the cattle out of some fields in times of surplus and storing the grass as either *hay* or *silage* for use in times of shortage.

Hay and silage

There are several different ways of making both hay and silage, but in each case the principle is the same.

A hay barn

Hay making

Hay

Hay is sun- and wind-dried grass. It does not rot away as there is not enough moisture for the organisms of decay (fungus and bacteria) to survive.

Silage

Silage is made by excluding air from fresh grass. The organisms of decay start the rotting process, but cannot continue owing to lack of air. Organisms which can live without air take over the decaying process, but these produce an acid which kills them off. The acids 'pickle' and so preserve the grass.

Silage making. A forage harvester lifting grass in a self-unloading trailer

Grass being sealed with polythene

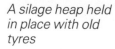

A silage heap held in place with old tyres

The next experiment tests the idea that air is needed for grass to decay.

Experiment 12.1
(*best time May or June*)

Items required: large plastic bag; lawn clippings; cardboard box; vacuum cleaner

1. Get two stándard wheelbarrows full of *fresh* grass. (Lawn clippings may be used but taller, more mature grass gives much better results.)
2. Make several holes into each side of a cardboard box and put in one barrow of grass.
3. Put the other grass into a large strong plastic bag.

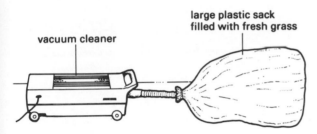

large plastic sack filled with fresh grass

vacuum cleaner

4. Use a vacuum cleaner to draw as much air from the bag as possible.
5. Seal the bag.
6. Leave for 4 weeks.
7. Examine the contents of both the box and the bag and compare. Some farmers use plastic bags to make silage (see photo).

Hay or Silage?

Each farmer has to decide whether to make hay or silage with surplus grass. There are many things to consider when making this decision:

	Hay	Silage
Weather	At least 3 dry sunny days are needed	Only 1 dry day needed
	The crop spoils if it is wet for a long time	Unlikely to be spoilt by the weather
Yield	One crop each year	Two crops each year
Making	Fully mechanised	Fully mechanised
Feeding	Light and easy to handle	Wet, heavy and difficult to handle – self-feeding systems required
Nutrition	Good	Very good – but unsuitable for baby calves and horses
Pollution	No pollution	Effluent (waste liquid) a problem
Health	Dusty hay can cause cancer	No problem – moist, no dust

If you had the choice to make – which would you choose?

Summary

1. More grass is grown in the UK than any other crop.
2. Grass can withstand grazing as its growing parts are at the base of the plant.
3. A *perennial* is a plant which flowers and seeds each year for a number of years.
4. A crop of grass which lasts for only a few years is called a *ley*. One which lasts many years is called *permanent pasture*.
5. Most pastures contain clover plants as well as grass plants.
6. *Hay* is grass which is preserved by sun drying.
7. *Silage* is 'pickled' grass.

Questions

1. Using words and diagrams, explain why each of the following plants is not destroyed by the grazing animal:
   (a) grass;
   (b) clover;
   (c) nettle.

2. (a) Describe two different methods of preserving grass for feeding to livestock in winter.
   (b) Give one advantage and one disadvantage of preserving the grass in each of the methods you mention.

3. Bales of hay have an average mass of 25 kg. How many bales would be required to:
   (a) Load a 5 tonne trailer to capacity? (1 tonne = 1000 kg)
   (b) Feed a herd of 50 cows for 10 weeks if each animal needs 12.5 kg a day?
   (c) Feed a pony for 140 days if the pony eats two bales a week?

Unit 13
Goats

Goats are *herbivores*. We use them to convert plant material into food (milk and other dairy products) and fibre.

Food

A goat feeds by *browsing* not *grazing*. A browsing animal takes the tips of shoots from any vegetation it can reach including hedges, bushes and trees. A grazing animal (e.g. cattle and sheep) crops the shorter vegetation from the ground. In a broadleaf woodland, where there is plenty of open canopy, a grazing animal would get most of its food from field-layer plants. The browsing animal would get food from the field layer, the shrub layer and the tree layer as far as it could reach (see Unit 22).

Goats will eat almost any weed, shrub or tree, including flowers and fruit. In spring and summer a goat with a good browsing range will find all the food it requires. This is a full-time job for the goat as fresh green food contains a lot of water; large amounts have to be eaten to supply all the goat's requirements. Some green food is poisonous and many goats will avoid it, or eat only very small amounts. However, where green food is collected and carried to housed animals there is a danger that if poisonous plants are offered they may be eaten.

Practical 13.1

1. Take a book on common wild flowers and trees to help you to identify the plants.
2. Collect a leaf and a flower (or fruit) from as many of the ten plants listed in the 'poison box' as you can find.
3. Mount your collection on cards and care-

Safety note: remember these plants are poisonous – do not put them near your mouth, and wash your hands after touching them. Wear gloves if you have sensitive skin.

fully cover them with the type of sticky-back plastic which is used to protect books.
4. Make a display of your collection with a card by each one to show the effect it would have if fed to a goat.

Poisonous plants

Plant	Effect on the goat
Cherry laurel	Sickness, pain, scour
Rhodo- dendron	Sickness, pain, scour
Yew	Sudden death
Laburnum	Sickness, large pupils, coma
Foxglove	Quick breathing, heart failure
Ragwort	Liver damage – no cure
Mare's tail	Scour, poor condition
Nightshades	Stomach upset, death
Potato leaves and green potatoes	Stomach upset, death
Bracken	Blood in droppings, death

Note: An animal is said to 'scour' if it produces very watery droppings.

During the autumn, winter and early spring goats will not be able to harvest enough food to keep themselves in good condition. They must be given extra food in the form of hay, roots and concentrates.

Hay

Hay is the most important food in the winter ration. Good dust-free hay should always be available in a rack above the floor as it is essential that the goat has enough fibre to fill its large *rumen* (i.e. its first stomach – a goat has four stomachs).

Roots

Carrots, turnips, swedes and mangolds can be fed whole or chopped. Before chopping, any soil on the roots must be washed off. Goats will not eat dirty food.

Concentrates

These can be purchased as 'goat nuts' or 'goat mixture'. They contain wheat, barley, oats, maize, bean meal, fish meal, soya meal, linseed, etc., plus minerals and vitamins. The trough in which concentrates are fed must be kept very clean. In common with other animals, a goat must always have access to *clean*, fresh water.

Housing

Goats become distressed if they get wet. They must have access to a dry, draught-free house with good ventilation. A warm, clean, dry bed is also essential. The goat must also have plenty of

The points of a milking goat

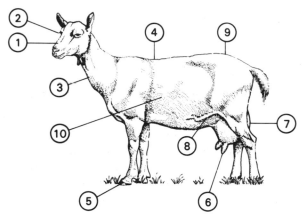

1 Head carried in an alert manner
2 Eye bright with a gentle look
3 Long fine neck
4 Long level back
5 Feet sound with hooves well trimmed
6 Teats placed well apart, pointing forward
7 Both sides of the udder a similar size
8 Large milk vein
9 Broad pelvis
10 Large deep body with a good spring of ribs

room to move around. Never attempt to keep a goat in a pen where it cannot get exercise; not only is this cruel but small pens are difficult to keep clean.

The next practical looks at how much time goats spend moving around, eating, sleeping and doing other things.

Practical 13.2

1. Copy the chart below:

Activity	Time (min):	1	2	3	4		18	19	20
Eating									
Walking									
Running									
Browsing									
Lying									
Standing									
Cudding									
Grooming									

2. Put a small amount of food outside and release the goat(s).
3. Select one animal and start a stopwatch.
4. Observe your animal for 20 minutes. After every minute, tick the box, or boxes, to show your animal's activity during that minute.
5. Record your results on a bar chart.

Breeds

A *breed* is a group of animals which have similar *points*. A breed is developed by mating animals which show a certain quality, and avoiding mating the ones which do not have it. If, for example, a breeder wanted only white animals then only white animals would be used for breeding. Other colours might be kept but would not be bred from. After several generations of selective breeding, the young of these animals would be all white. At the same time, other features (like milk yield, docile behaviour, size, etc.) will be selected; when a group of

50% GOATS WHITE – ONLY WHITE ANIMALS USED FOR BREEDING

3 years later:

ONLY WHITE ANIMALS USED FOR BREEDING

6 years later:

ONLY WHITE ANIMALS USED FOR BREEDING

9 years later:

The goats are now all white. *Note:* some characteristics would take much longer to breed in; also the breeder is attempting to breed for a number of different characteristics at the same time

Anglo-Nubian goat

SAANEN BREED

TOGGENBURG BREED

animals have these features consistently they can then be called a breed.

It is possible to increase the number of breeds in this country, either by breeding new ones or by importing a new breed from abroad. The

Cashmere and Angora breeds of goats have recently been imported. These breeds are kept for their coats.

Common breeds are:

Anglo-Nubian: These are easily recognised by the long, floppy ears and the Roman nose.

Saanen: These were originally imported from Switzerland. The breed is white in colour, hornless and very tame. Saanens are very good milkers and may give up to 6 litres a day.

Toggenburg: These are slightly smaller than the Saanen. They are fawn or light brown with white markings. Some adults grow long hair along the back and down the back legs.

Breeding

Season

A female goat will not allow the male to mate unless she is *in season*. The season lasts from 1 to 3 days and occurs once every 3 weeks.

A female in season will make loud and regular calls. She will shake her tail, be rather restless and her vulva will be red and swollen. Females come into season before they are fully grown at about 9 months old. They should not be mated, however, until they are 18 months old.

Kids

The female will produce a kid about 150 days after mating. This length of time is known as the *gestation period*, and it may vary by a few days. If the goat is giving milk whilst she is pregnant she should be dried up at least 8 weeks before kidding. This will give her time to build up her body reserves ready for the birth and lactation (*lactation* = giving milk).

After kidding the female will lick her young. This stimulates the kid into life and helps to form a bond between mother and baby. The mother may then lie down and have a second kid. After the last kid has been born the *afterbirth* appears. This consists of the membranes which held the kids whilst they were developing inside the uterus (*uterus* = womb). The afterbirth should be removed and buried before the mother attempts to eat it.

The kid(s) needs to suckle its mother as soon as possible. The first milk produced after giving birth is known as *colostrum*; it contains special substances called *antibodies* which protect the young against disease. If the kid does not suckle, help it to do so by squeezing a little milk from teat to mouth. After the first two or three feeds the kid can be trained to drink its mother's milk from a bottle, or it can be left to suckle naturally.

Milking

The female goat should be milked twice each day into a thoroughly cleaned and sterile container. Milk is then strained (filtered) through a milk filter and cooled immediately with cold running water. You will find that goat's milk does not keep fresh as long as purchased cow's milk as it has not been *pasteurised*. (This is a heat treatment which kills most bacteria in the milk. It is the presence of bacteria which turns milk sour.)

Milk

Practical 13.3 Goat's milk

A. Microscopic examination

1. Take a fresh sample of goat's milk. Place a single drop on a microscope slide.
2. Cover with a cover slip. If any bubbles of air appear under the cover slip, try again.
3. Place under a microscope and view on *low* magnification.
4. Turn to a higher magnification and view again.
5. Can you see any movement? Using compasses, draw a circle in your note book.
6. Inside the circle draw all you can see down the microscope.
7. Repeat this using cow's milk. Are there any differences?

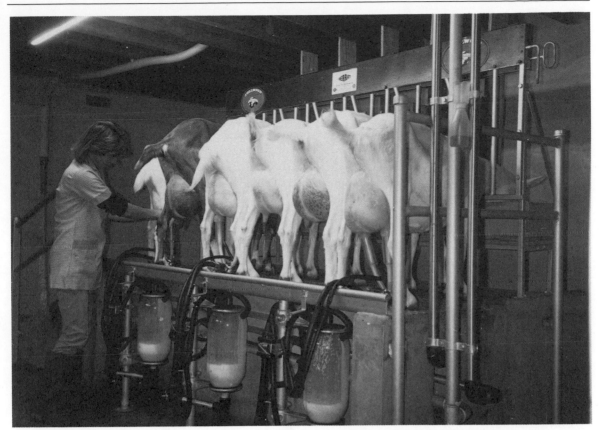

A milking parlour

B. To find its density

This is easily done with a special hydrometer known as a lactometer. If you do not have one use the following method:

1. Weigh a measuring cylinder and record the mass.
2. Take two beakers, one of water and the other of goat's milk, and use a thermometer to check that both milk and water are at the same temperature.
3. Pour exactly 100 ml of the water into the measuring cylinder and weigh. Record the mass.
4. Tip away the water and dry the measuring cylinder.
5. Pour exactly 100 ml of the milk into the measuring cylinder and weigh. Record the mass.
6. Calculate the mass of water and the mass of milk.

7. Calculate the density of goat's milk using this formula:

$$\text{density} = \frac{\text{mass of goat's milk}}{\text{mass of water}}$$

8. Repeat this for cow's milk. Are the densities the same?

C. Finding out what it contains

We are told that cow's milk contains *fat, sugar* and *protein*. Use the following three tests to see if goat's milk contains the same food substances:

(i) Sudan III test for fat

1. Add the juice of half a lemon to 250 ml of warm (40 °C) goat's milk. Five minutes later a curd will appear in the milk.
2. Take a little curd and press it on to clean filter paper. Remove the curd.
3. Add a few drops of sudan III. Leave for 2 minutes and then wash off surplus dye with distilled water.

4. If fat is present the dye turns red.
5. Rub a little curd between your finger and thumb. Does it feel greasy?

(ii) Protein (Biuret test)

1. Pour 6 ml of goat's milk into a measuring cylinder.
2. Add 4 ml of 5.0 mol dm^{-3} sodium hydroxide to the cylinder.
3. Pour the mixture into a test-tube.
4. Carefully add 0.05 mol dm^{-3} copper(II) sulphate solution, drop by drop, until the contents of the test-tube are coloured.
5. A precipitate will begin to form. A pink/ purple-coloured precipitate indicates protein; a blue colour indicates no protein.

(iii) Sugars (Fehling's test)

1. Pour 6 ml of goat's milk into a measuring cylinder.
2. Add 2 ml of Fehling's solution A and 2 ml of Fehling's solution B to to the milk.
3. Pour the mixture into a test-tube.
4. Heat the tube very gently by passing it backwards and forwards through a low bunsen burner flame.
5. If sugar is present a brick-red colour will appear. If there is no sugar the solution will not change colour.

Yogurt

Yogurt is formed from milk by the action of special *bacteria* which are quite harmless. Natural yogurt on sale in shops usually contains some of these bacteria. A little purchased natural yogurt can be used as a 'starter' to convert other milk into yogurt.

Test to see if the bacteria which convert cow's milk to yogurt will also convert goat's milk to yogurt.

1. Immerse all the apparatus you are going to use in boiling water to kill any bacteria which may be present (i.e. sterilise it). (Make certain that the thermometer you are going to use will measure temperatures above 100 °C, or it will break.)
2. Pour 500 ml of goat's milk into a saucepan.

Bring it to the boil. (Take care!)
3. Allow it to cool, noting the temperature from time to time.
4. When the temperature reaches 43 °C, add one tablespoon of natural yogurt and 25 ml of milk powder.
5. Mix it well and pour into a vacuum flask. Replace the flask top.
6. Leave undisturbed for 8 hours.
7. Stand a basin in cold water and pour in the milk. Stir gently to speed up cooling.
8. Cover the basin and leave in a refrigerator for 3 hours.

Making soft cheese from goat's milk

1. Warm 500 ml of goat's milk in a clean saucepan and warm to 38 °C.
2. Add 40 ml of lemon juice. Leave for 15 minutes and observe what happens.
3. Line a sieve with moist muslin.
4. Pour the contents of the saucepan through the sieve.
5. Tie the corners of the muslin together and hang it up to drain. Leave it hanging for half an hour.
6. Remove the cheese from the muslin.
7. Add a little salt; mix and taste.

Junket

A goat kid's stomach produces a substance which causes the milk it drinks to clot. This slows down the milk's passage through the gut and aids digestion. The clotting substance is a chemical which scientists call an *enzyme*; this one is *rennin*. Rennin which has been extracted from the stomach of calves is available from chemists shops.

1. Purchase some rennet. Attempt to make junket from goat's milk instead of cow's milk by following the instructions on the packet.
2. Attempt to make a number of smaller quantities of junket using temperatures both above and below the recommended temperature. (say 20°, 30°, 50°, 60° C). What difference does the temperature make?

Butter

1. Stand a litre of goat's milk overnight.
2. Remove the top 40 ml.
3. Half fill a test-tube with some of the 40 ml you removed.
4. Cork the tube and shake for 3 or 4 minutes.
5. Observe from time to time.
6. If it is possible to make goat's milk butter then lumps of butter will begin to separate out.
7. Wash any lumps in water, salt and taste.

Summary

1. A goat is a *browsing herbivore*.
2. Browsing animals take food from ground-layer plants, shrub-layer plants and tree-layer plants.
3. Some plants are poisonous to goats.
4. Goats must be well housed.
5. A *breed* is a sub-set of a species, the breed having a smaller number of characteristics than the species.
6. Common breeds of goats are: *Anglo-Nubian, Saanen, Toggenburg, Angora, Cashmere*.
7. A female goat will mate only if she is *in season*.
8. A female goat should not be bred from until she is 18 months old.
9. The first milk a female goat produces after giving birth is a substance called *colostrum*.
10. Colostrum is an essential first feed for kids as it contains *antibodies* which protect the young from some diseases.
11. Milk contains *water, sugar, fat* and *protein*.
12. Some *bacteria* which are present in milk turn it sour.

Questions

1. Write single sentences to answer the following questions:
   (a) How does grazing differ from browsing?
   (b) What is rennet extracted from?
   (c) Why is it necessary to add a little yogurt when making yogurt?
   (d) Which breed of goat has a Roman nose and floppy ears?
   (e) Why have the Angora and Cashmere goats been imported into this country?
   (f) What are concentrates?
   (g) What is a lactation period?
   (h) Name three plants which may cause a goat to scour.

2. Draw a pie diagram to show the substances which are present in goat's milk. Make the angles as follows:
   protein 15, sugar 19, fat 23 and water 303

3. You have a milking goat and are going away on holiday. A friend agrees to look after your goat for 2 weeks while you are away. Write a list of instructions telling your friend exactly what he/she is to do.

4. (a) Describe an experiment you have carried out using goat's milk.
   (b) What did you learn from the experiment?

Unit 14
The earthworm

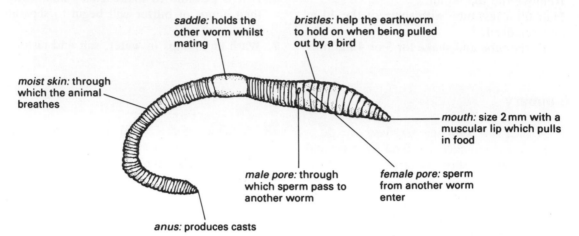

saddle: holds the other worm whilst mating

bristles: help the earthworm to hold on when being pulled out by a bird

moist skin: through which the animal breathes

mouth: size 2 mm with a muscular lip which pulls in food

male pore: through which sperm pass to another worm

female pore: sperm from another worm enter

anus: produces casts

The external features of an earthworm

Earthworms are of very great importance in maintaining the soil, removing dead plant material and providing food for other animals:

1. They make *tunnels* in the soil:
   (a) Tunnels fill with air – this is needed by plant roots for respiration.
   (b) Tunnels help water to drain away from the surface.
   (c) Tunnels have leaves pulled into them from the surface. The leaves decay into plant food.
   (d) Tunnels provide spaces in which roots can grow.
2. Earthworms eat large amounts of fine soil both with their food and as they dig tunnels. This soil is deposited with droppings called *casts*. These are usually deposited on the surface of the soil. This covers the surface of the soil with a fine material suitable for germination and root growth.
3. Earthworms remove dead leaves from the surface. If leaves were not removed they would lie on top of the soil and form a layer of peat. Many small plants would be unable to grow through this.
4. Earthworms form an important part of the *food web* as they are eaten by moles, foxes, badgers, otters, wood mice, shrews, thrushes, blackbirds, crows, robins, owls, rooks, gulls, lapwings, starlings and many more.

The next experiment tests the idea that earthworms eat dead leaves, but not those with a strong smell.

Experiment 14.1

Items required: large plant pot; soil; earthworms

1. Put moist topsoil loosely into a 22 cm pot to about 5 cm from the top.
2. Dig up six large earthworms (handle with care).
3. Place the earthworms on the surface of the soil and observe them as they burrow into the soil.
4. Place small pieces of leaves on the surface. Use the following: cherry, carrot, celery, cabbage, mint, eucalyptus, onion and rue.
5. Cover the pot to prevent other organisms or wind from removing the leaves and place it outside. (Your classroom is probably too warm for worms, which live in a cool environment.)
6. Inspect it daily and replace any leaves which disappear.
7. Keep careful records of your observations.

Practical 14.1

(*best done in the autumn*)

Items required: watering can; spade; paint scraper; balance

1. Find an unused corner of a freshly dug garden plot.
2. Wet it with a rose can. At the same time pat it gently with the back of a spade until the surface becomes 'capped' with a thin layer of mud.
3. Peg out 1 square metre which can easily be reached from the path.
4. Each morning, using a paintscraper, collect any wormcasts. Dry them, weigh and record.

In 1881 the great scientist Charles Darwin estimated that each year earthworms left casts totalling over 2 kilograms for each square metre of soil. He also suggested that the activities of earthworms were responsible for burying Roman remains. (There are, however, other reasons why these get buried.) Charles Darwin also carried out an experiment like the one you have just done and formed the idea about their food which you have just tested.

Breeding

The earthworm has both male and female organs. When mating, both worms give and receive sperm.

Some time after mating each earthworm produces an oval-shaped cocoon from the saddle. The cocoon contains eggs in a nutritious fluid. After a month or more the eggs hatch into miniature worms. These baby earthworms take about a year to grow large enough to produce their own young.

Practical 14.2

Items required: wormery; peat; sand; compost; lime; leafmould

If you do not have a wormery you can make one from an old plastic ice cream container and two pieces of glass as in the diagram.

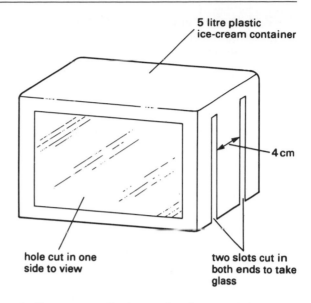

hole cut in one side to view

5 litre plastic ice-cream container

4 cm

two slots cut in both ends to take glass

1. Tape a small piece of soft material to one end of a 30 cm rule to act as a duster.
2. Put a 3 cm layer of leafmould at the bottom. Level it with one end of the ruler and carefully clean the glass with the duster at the other end.
3. Sprinkle a *small* quantity of lime over the leafmould to help mark the boundary between the leafmould and the layer above.
4. Repeat parts 2 and 3 above for the other materials until the wormery is full to within 5 cm of the top.

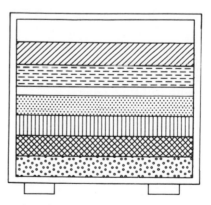

The completed wormery

5. Dig up three large earthworms and place them gently on to the surface.
6. Cover the sides of the wormery to exclude the light, as it is harmful to earthworms.

7. Keep the material moist by adding a little water as necessary.
8. Uncover and observe every 2 or 3 days.
9. Record your observations with a series of diagrams.

Whilst digging up earthworms with a spade you may chop one in half. It is possible that this will kill the worm. It is also possible that the front part of the worm will live and its tail will heal. The cut-off tail portion will certainly die.

Another method of collecting earthworms is as follows:

1. Measure 25 ml of 40 per cent formaldehyde and pour it into 5 litres of water. (Do not make the mixture any stronger or you will harm the worms.) A solution of potassium permanganate is sometimes used for this purpose, but it is better not to use this as it also poisons the worms.
2. Put the mixture into a watering can and fit a rose.
3. Walk to the sampling area treading lightly (vibrations may make the worms descend deeper into the soil).
4. Soak about half a square metre of soil and the worms will come to the surface.
5. Use forceps to pick up the worms and briefly wash them in cold water.

This method (formaldehyde) is sometimes used to measure the populations of earthworms present in soils. The results from these counts show that earthworms do not live in very acid soils. They also show that there are very many more earthworms under pasture (a greater mass than the total mass of the animals which are grazing the pasture) than there are in cultivated fields. The reasons for this are that cultivated soils are more likely to dry out than soils covered with grass and that predators can get at worms more easily when there is no turf to protect them. (Flocks of gulls near to a working plough are generally feeding on earthworms.) Earthworms often come to the surface to feed at night. They are then prey for the fox, owl and other hunters.

Summary

1. The earthworm is important as it keeps the soil in good condition by aerating it, mixing it up and pulling leaves down into it.
2. The earthworm is an important food for many creatures.
3. Earthworms produce *casts* of fine soil on the surface of the ground.
4. Earthworms have *both* male and female sex organs and after mating *both* worms lay eggs.
5. Earthworm eggs hatch into tiny worms.
6. Drying out and light are both harmful to earthworms.

Questions

1. **Explain in detail what the wormery taught you about the probable effects earthworms have on soil.**

2. (a) **Draw a large diagram of an earthworm. Label the parts.**
   (b) **Draw a series of diagrams to show how the earthworm moves when on the soil surface.**

3. (a) **Describe how it would be possible to estimate the number of earthworms which are present in the school sports field.**
   (b) **How would you expect these numbers to differ from the numbers in the school garden?**
   (c) **Give reasons for your answer to 'b' above.**

Unit 15
Soil fauna

Soil fauna (soil animals) are very important for the following reasons:

1. They break down unwanted *organic matter* (like leaves and animal droppings) and return it to the soil.
2. Their droppings provide food for *bacteria* and other microorganisms which are essential in a healthy soil.
3. They make *tunnels* in the soil which fill with air; this allows animals and plant roots to breath.
4. Some soil animals eat our crops.
5. Some soil animals feed on pests.

Soil animals

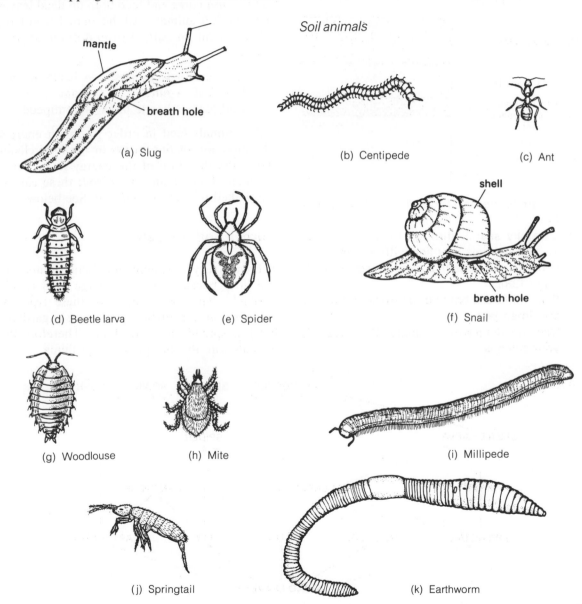

mantle

breath hole

(a) Slug

(b) Centipede

(c) Ant

(d) Beetle larva

(e) Spider

shell

breath hole

(f) Snail

(g) Woodlouse

(h) Mite

(i) Millipede

(j) Springtail

(k) Earthworm

Practical 15.1 Becoming familiar with different types of soil animals

Items required: leaf litter; pooter; plastic teaspoon; collecting jars

The leaf litter underneath deciduous trees contains many different species of soil animals.

1. Collect about 20 litres of leaf litter from under deciduous trees. Take material from each layer, including that which is in contact with the soil surface.

Layers in leaf litter

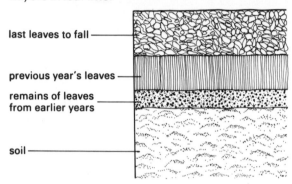

last leaves to fall

previous year's leaves

remains of leaves from earlier years

soil

2. Return to the classroom. Tip the litter on to a table.
3. Carefully search through for animals. Pick up any animals you see with a plastic teaspoon or (if small) a *pooter* (see Unit 25 page 138).
4. Place different types of animals in different specimen jars.
5. Use the diagrams on page 81 to identify your catches.
6. Count the numbers of each type, and record in your notebook.
7. Return the animals to the litter and then place it on the compost heap.
8. Display your results on a bar chart.

Food chains and food webs

Not all the animals you have collected will be *herbivores* and feed on dead leaves. Some will be *carnivores* and feed on other animals. Some will be *omnivores* and feed both on dead leaves and on other animals. All the animals you have collected will be part of several different *food chains*:

dead leaf → woodlouse → beetle larva
dead leaf → mite → ant
dead leaf → beetle larva → centipede

All animals feed in order to obtain energy. The direction of the arrows in the food chains shows the direction of the energy flow. There are many food chains in the soil; these can be interlinked to form a *food web*. See below.

Garden soil animals

There are very large numbers of small animals in leaf litter because there is a large food supply available. The food supply for these types of animals is also available in soils (e.g. a garden), but it is spread out more thinly. Therefore the animals will also be spread more thinly.

A food web

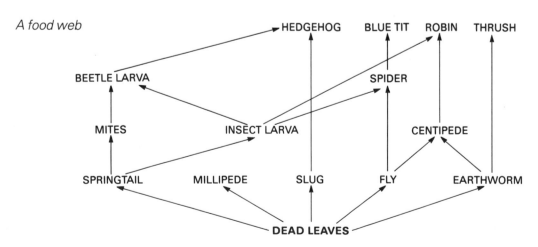

Practical 15.2 To find what soil animals are present in a garden plot

Items required: grapefruit peel; turnip; apple; plastic cup; plastic sheet; trowel; collecting jars

1. Set up each of the following collecting methods on a single garden plot 1 week before collection:
   (a) Place a 60 cm by 60 cm concrete slab on a level area of soil. (Take care! – let your teacher supervise and you will be shown how to lift it without hurting yourself.)
   (b) Rake a hollow area 4 cm deep and 1 metre square. Lie a metre square of black polythene on the bottom and cover it with soil.
   (c) Fill a plastic cup with short, dry grass and bury it, lying on its side, so that the top is just 1 cm below the surface.
   Set up the following 1 day before collection:
   (d) Place the peel of half a grapefruit on the soil.

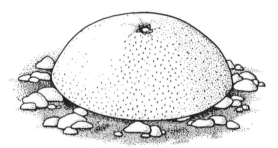

   (e) Bury a plastic cup in the soil so that the lip is just level with the surface. Protect the cup from rain and birds with a large

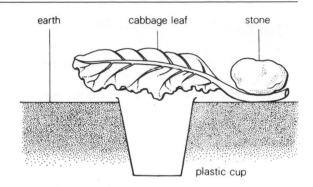

leaf or piece of plastic, but leave enough room for invertebrates to walk underneath.
   (f) Repeat part (e) but fill the cup with a mixture of half beer, half water and a teaspoon of brown sugar.
2. On the day of collection, remove and search all your traps. Put the animals caught in a plastic container with a lid in which there is a handful of dead leaves (a 2 kg margarine container is ideal). Dig and search in the top 20 cm of soil and add any animals you find to the collection.
3. Take your catch back to the classroom. Identify and count your specimens before returning them to the soil.
4. Record your results on a bar chart. Compare this with the bar chart you made after the leaf litter search. Can you explain any of the differences?
5. Construct a food web. Include in it the shrew (there are almost certainly some nearby), the hedgehog, thrush and blackbird.

Summary

1. Soil animals (*soil fauna*) break down *organic matter* and return it to the soil.
2. Soil animals make *channels* in the soil which fill with air.
3. Some soil animals are crop pests; others feed on the pests.
4. Some soil animals are *herbivores*, some are *carnivores* and some are *omnivores*. (If you do not understand these words look them up in the Glossary on pages 152–155.)
5. In a *food chain* the arrows indicate the direction in which the energy flows.
6. Food chains can be interlinked to form a *food web*.
7. Soils contain similar animals to leaf litter but the animals are more widely spread.

Questions

1. Write single sentences to answer the following questions:
   (a) Why is the presence of air in soil important?
   (b) In what ways does a centipede differ from a millipede?
   (c) Why does a pitfall trap have to be covered?
   (d) What does a carnivore feed upon?
   (e) What happens to the droppings of soil animals?
   (f) What do the arrows in a food web indicate?

2. Construct a food web containing all the animals which are illustrated in this Unit, plus the hedgehog, the mole, the shrew and the thrush.

3. Samples of the soil animals were taken from two different areas. Area 'A' was a wet area where the soil was clay. Area 'B' was a dry area where the soil was sand.

	Area 'A'	Area 'B'
Mites	68	78
Spiders	6	18
Woodlice	42	14
Ants	0	32
Centipedes	12	16
Millipedes	16	9
Slugs	35	6
Beetles	23	35
Harvestmen	6	17

(a) Draw a single bar chart to include all this information.
(b) Which soil had the greater numbers of animals?
(c) Which animals are best suited to a wet, clay soil?
(d) Which animals are best suited to dry, sand soil?

Unit 16
Cowpats

Each cow produces 10 or 12 large cowpats a day. If these did not disappear they would soon cover the pasture and there would be no grass to eat. The cow does not extract all the energy from the grass she eats; there is still plenty left in the cowpat for other creatures to make use of. A cowpat quickly becomes home for hundreds of creatures. A community builds up which contains both herbivores and carnivores. Some of these creatures are large enough to attract bigger animals such as foxes, badgers and several birds. The end result is that in summer the cowpats quickly disappear. In the colder days of winter they disappear much more slowly. Rain, wind and frost crumble and decay them.

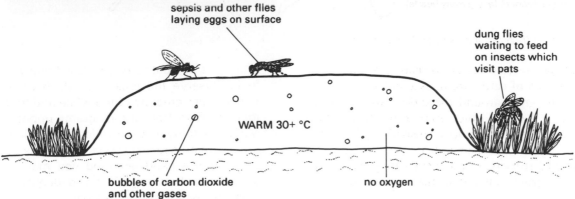

Section through a cowpat – 10 minutes old

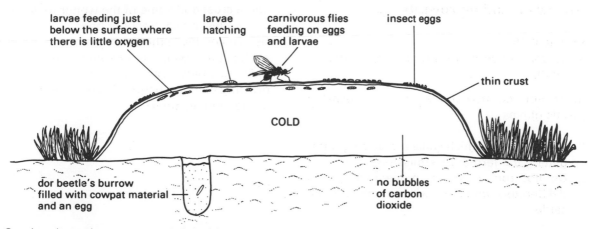

Section through a cowpat – 1 day old

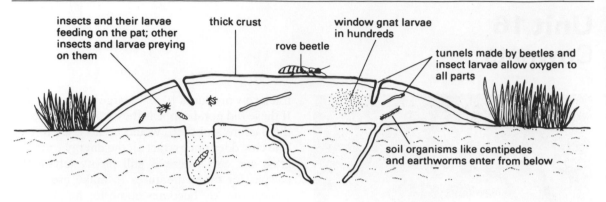

Section through a cowpat – 1 week old

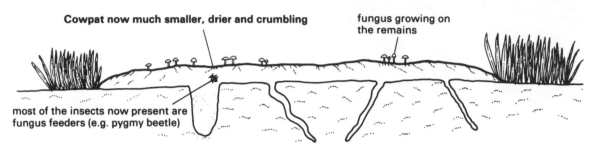

Section through a cowpat – 2 weeks old

There are also fewer cowpats in winter as the majority of cattle are housed indoors.

The grass around the older pats is a darker green than the other grass in the field. This is because the substances released by the cowpat animals get washed into the soil and they are good fertilisers. Cattle avoid eating this grass; it soon grows taller than the surrounding grazed area.

Animals found on cowpats

Practical 16.1

(*best done in summer term or September in hot sun*)

Items required: trowel; plastic bag; disposable plastic gloves

1. Visit a field in which cattle are grazing (after getting the farmer's permission).
2. Do not run or shout; move quietly and deliberately and you will not disturb the cattle.
3. Wait for a cowpat to be produced and walk to it. Observe the creatures which arrive. Watch these creatures for a while and try to identify them from the diagrams opposite.
4. Move to a fairly new pat which has a skin on top. Can you see any insect eggs? Observe these with a hand lens and you may see some larvae hatching out. Try to identify the creatures which are on this pat.
5. Find an older pat which has a crust, covering wetter material. Use a trowel to put some crust and some of the wetter material into a plastic bag. Seal the bag.
6. Before you leave the field have a good look at the grass which is growing around the older cowpats. Can you see any fungus growing on these pats? Can you identify any of the animals on it?

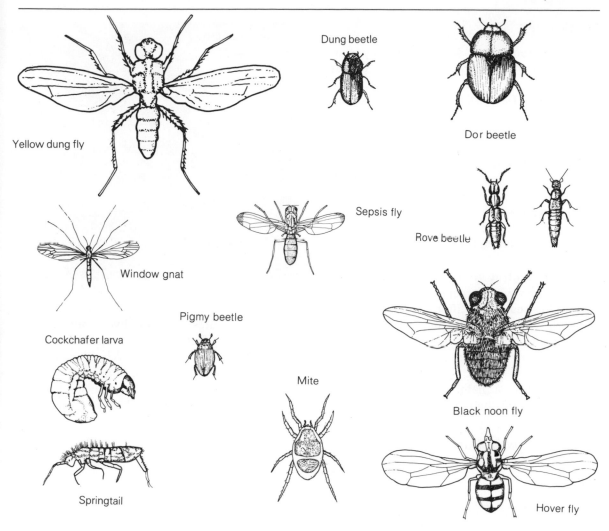

Yellow dung fly

Dung beetle

Dor beetle

Window gnat

Sepsis fly

Rove beetle

Cockchafer larva

Pigmy beetle

Mite

Black noon fly

Springtail

Hover fly

Dung fly: This visits very fresh pats to prey on insects and to lay eggs in the pat. It is a carnivore. The larvae feed on cowpat material.

Dung beetle and its larva: The larvae feed on cowpat material.

Dor beetle: This digs a hole through the pat and into the soil. It fills the soil hole with cowpat and lays an egg. After the egg hatches the larva feeds on this store. The adults also feed on dung.

Window gnat: The larvae look like very thin worms (2–4 mm long); they are usually present in large numbers, feeding on the cowpat material. The adults are herbivores.

Sepsis fly: This is easily recognised by the way it holds its wings – straight out and quivering. It is an early coloniser, laying eggs on fresh pats.

Rove beetles: There are several different kinds

all of which are carnivores; they have long bodies with short wing cases.

Beetle larva: There are many different beetles; the young larvae eat cowpat material and become carnivorous as they get older.

Pigmy beetle: Only 1 mm long – it is found only in older pats as they feed on fungi.

Black noon fly: This is the largest fly to visit. It lays only one egg in a single cowpat. The larva is similar in appearance to a fishing maggot, and is the largest larva present. They feed on dung; the adult sips nectar from flowers.

Springtail: This is a herbivore. There is no larval stage; the young look like small adults.

Hover fly: The larva is a carnivore.

Mite: These tiny eight-legged creatures feed on fungus in old cowpats.

Practical 16.2

Items required: metal sieve; white tray; hand lens; petri dishes; forceps, disposable gloves

1. About half fill the sieve with cowpat material.
2. Using an outside tap, run water through the material.
3. Stir the material gently with a gloved hand and continue washing it through.
4. Repeat part 3 above until the water running from the sieve is clear.
5. Return to the classroom and empty the sieve into a white tray which has about 4 cm of water in it.
6. Sort through it with care; pick out all the organisms with forceps and place them in petri dishes. Use different dishes for different types of organisms.
7. Count the organisms in each petri dish.
8. Enter your results on a bar chart.

The cowpat is a habitat for many different creatures. As this habitat soon disappears most of these creatures are short lived, but in their short lives they feed, grow and reproduce. They also play a vital role in the cycle of nature by turning cowpats into food for larger animals and returning vital substances to the soil, which plants need in order to grow.

Recycling of materials through a cowpat

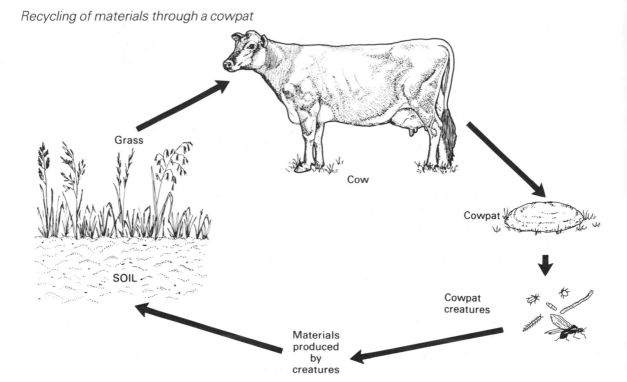

Summary

1. Cowpat material contains *energy* in the form of undigested food.
2. Very fresh cowpats contain a lot of *carbon dioxide*.
3. Creatures which inhabit cowpats are usually short lived.
4. Many of the cowpat creatures are *carnivores*.
5. Cowpats are also broken down by *fungi*.
6. Cowpat organisms are food for larger animals and birds.
7. Materials from broken-down cowpats enter the soil and are used by plants.

Questions

1. **A sample of 5-day-old cowpat material was washed out and the following creatures were found:**

rove beetles	14	ground beetles	7
dung beetle larvae	10	black noon fly larva	1
window gnat larvae	32	dung beetles	5
sepsis flies	26	unidentified	18

   Display these results on a bar graph.

2. **Explain with words and diagrams why pasture fields do not become completely covered with cowpats during summer grazing.**

Unit 17
Weather recording

The main things which make up our weather are:

1. Temperature.
2. Sunshine.
3. Wind.
4. Precipitation (rain, hail and snow).
5. Humidity (including mist and fog).
6. Cloud cover.
7. Air pressure.

To this list the rural scientist must add soil temperature. This is very important as plants will not grow if their roots are cold. It is the temperature of the soil which determines the start of spring.

Temperature

The ordinary Celsius thermometer, which is filled with either mercury or alcohol, will give the temperature at any time of day or night but will not record it. The best thermometer for the rural scientist to use is the *maximum and minimum thermometer*. This records the highest and the lowest temperatures reached since it was last set. The thermometer is usually set each day, so it gives the highest day temperature and the lowest night temperature in the previous 24 hours.

The maximum and minimum thermometer has mercury in one part of the tube, and alcohol at both ends. The alcohol flows around two iron riders, but the mercury pushes them along. They stay at the point where they were pushed to, showing how far the mercury has moved. One end of the mercury is marked to show the highest temperature; the other end is marked to show the lowest temperature.

To make the instrument more compact it is bent into a 'U' shape.

There are several ways of pulling the riders back down to the mercury. The most common method is to use a magnet.

(*Note:* The iron riders do not rust as there is no air or water inside the tube)

Practical 17.1 Recording the highest and lowest temperatures reached in a greenhouse

Items required: maximum and minimum thermometer

1. Move the riders down until they just touch the ends of the mercury thread.
2. Hang the thermometer just above the top of the plants in the centre of the greenhouse.
3. Twenty-four hours later read the temperatures which are level with the *bottoms* of the riders and record.
4. Reset the thermometer.

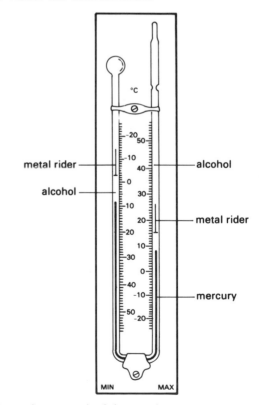

A maximum and minimum thermometer

Practical 17.2
(*This can only be done during sunshine*)

Items required: two maximum and minimum thermometers

1. Place the two thermometers outside in direct sunshine.
2. Cover one with a large cardboard box in which there are lots of holes to allow good air circulation.
3. Set both thermometers.
4. Leave for an hour.
5. Read both thermometers and record the temperatures.
6. Repeat the experiment on a wet day, making sure the box keeps one thermometer dry.

At weather stations thermometers are housed in a *Stevenson's screen*. (This is a white-painted box with slats, stood on legs. It is shown in the photograph.) Look at the results of your practical work and say why.

Stevenson's screen

Inside a Stevenson's screen

Task 17.1

The table shows temperatures taken during a 48-hour cloudless period at the author's school. One set of readings was taken at soil level; the other was taken in the Stevenson's screen 1 metre above it.

Draw two line graphs on one graph – one line to represent the temperature in the Stevenson's screen and the other to represent the temperature on the soil surface. Put the time along the bottom scale and the temperature up the side scale.

Time (h)	Temperature on the soil surface	Temperature in Stevenson's screen 1 metre above soil
00.00	15	15
06.00	20	16
12.00	42	25
18.00	28	28
00.00	13	13
06.00	20	15
12.00	45	25
18.00	33	25
00.00	14	14

Soil temperatures

Soil is a very poor conductor of heat. As a result of this the surface is usually a different temperature to the layers underneath.

The thermometer in the photograph is designed to measure the temperatures at a fixed depth (50 cm) in the soil.

A soil thermometer

Task 17.2

Examine the graph below and answer the questions:

1. What was the highest soil surface temperature?
2. What was the coolest soil surface temperature?
3. At what time of day is the soil on the surface the hottest?
4. At what time of day is the soil 20 cm deep the warmest?
5. What is the difference between the warmest and the coolest temperatures at each of the following depths?
   (a) At the surface;
   (b) 5 cm deep;
   (c) 10 cm deep;
   (d) 20 cm deep.

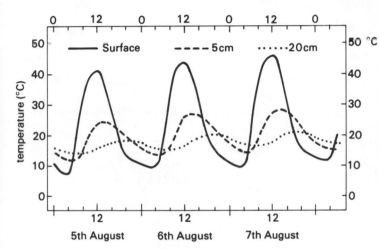

Use the graph and the figures you have calculated to describe the difficulty an earthworm is likely to have in the summer. (The highest temperature at which the earthworm is comfortable is 12 °C. Most of its food is in the upper 20 cm of the soil. The earthworm must also remain in a moist environment – see Unit 14.)

The temperatures of soil at different depths on three hot summer days

Frost

Frost occurs when the temperature falls below 0 °C. This is of great importance to the rural scientist as many plants are killed by frost.

During the winter when the soil freezes it is the top which is frozen. The deeper parts of the soil remain unfrozen. Our hardy plants are adapted to this and can survive these conditions. If these plants are grown in tubs, however, they may die. This is because the whole root ball freezes. Tubs which contain bulbs, shrubs and other plants should be protected from frost.

The frosts which cause most damage to our plants are not the winter ground-frosts but the late spring air-frosts. These frosts usually occur when the sky is cloud free (clouds keep the ground warm by trapping the radiation which would otherwise be lost into outer space). During these frosts the cold air sinks and lies in *frost hollows* and *frost pockets*.

Fruit trees and bushes (frost will damage the blossom) and half-hardy plants should not be planted in these pockets or hollows.

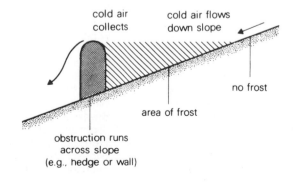

A frost pocket

A frost hollow

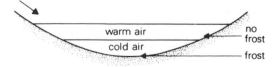

Sunshine

A sunshine recorder

The *sunshine recorder* focuses the sun on to a card at its rear and burns a hole in it. By measuring the length of the holes it is possible to tell how long the sun has been shining during the day. A new card must be fitted into this instrument each day.

Task 17.3

Look at the diagram of a sunshine recorder card. Calculate the period of sunshine.

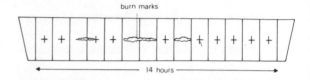

Wind

Two measurements have to be taken to find the effect of the wind:

1. Its direction; this is indicated by a *weather vane*.
2. Its speed; this is indicated by an *anemometer*.

A weather vane

The Beaufort scale of wind force

Beaufort number	General description of wind	Visual evidence of wind	Wind speed (m/s)
0	Calm	Smoke rises vertically	less than 0.5
1	Light air	Wind direction shown by smoke drift	0.5−1
2	Slight breeze	Wind felt on face; leaves rustle	1−3
3	Gentle breeze	Leaves and small twigs move; light flag extended	3−5
4	Moderate breeze	Raises dust and paper; small branches move	8−11
5	Fresh breeze	Small trees in leaf sway	11−14
6	Strong breeze	Large branches move; whistling in telegraph wires	14−17
7	Near gale	Whole trees sway; difficult to walk against wind	17−21
8	Gale	Breaks twigs off trees; generally impedes progress	21−24
9	Strong gale	Slight structural damage; chimney pots and slates blown off	24−28
10	Storm	Trees uprooted; considerable damage to buildings	28−35
11	Violent storm	Widespread damage (very rarely experienced)	over 35
12	Hurricane	Widespread destruction	

An anemometer

The weather vane points to where the wind is coming *from*. A south wind comes from the south and blows *towards* the north.

The direction from which the wind most often blows is called the *prevailing direction*. Over most parts of the UK the prevailing wind is south-west. As this wind has crossed the Atlantic Ocean it is humid; as it is coming from the south it is warm. When this wind rises over the mountains and moors in the west of the country it cools and drops some of its moisture as rain. This is the reason why there is more rainfall in the west than in the east of the country. It is also why grass is the main crop in the west.

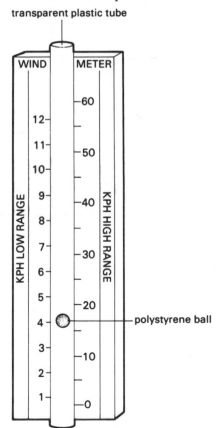

It is possible to estimate the wind speed by using the *Beaufort scale*.

The diagram shows a hand-held wind meter. As the wind blows across the top the ball rises. Try to make one yourself; use the Beaufort scale to graduate it. (A heavier ball can be used for the 'high range'.)

Task 17.4

Go outside and, using the Beaufort scale, estimate the speed of the wind. (You could use the meter below left.)

Precipitation (rainfall, snow, hail)

An automatic rain gauge

Rainfall is measured not by volume but by *depth*.

The *rainfall* is the depth of the lake which would form on a level surface if the rain did not soak into the ground or run off. It is measured in millimetres.

Task 17.5

Draw a bar chart showing the rainfall in your district in a single year.

If you do not have the figures use these (for Staffordshire):

Jan: 85 mm Feb: 112 mm Mar: 73 mm Apr: 63 mm
May: 18 mm Jun: 106 mm Jul: 35 mm Aug: 88 mm
Sep: 74 mm Oct: 45 mm Nov: 95 mm Dec: 79 mm

What was the total rainfall for the year?

Humidity

Humidity is the amount of water vapour which is present in the air. This vapour is not like steam from the kettle; it is an invisible gas. The humidity is measured as a percentage of the amount of moisture the air will hold. A high percentage (85+) is a humid atmosphere; a low percentage (70−) is a dry atmosphere.

Warm air will hold more moisture than cold air. So the temperature is important when measuring humidity. If the humidity reaches 100 per cent the air is saturated and cannot hold any more. At this high humidity water begins to condense into mist or rain.

Humidity is measured with a *wet and dry bulb thermometer*. In order to understand how this works try the following:
Wet one of your thumbs by putting it in your mouth and then hold both thumbs side by side. The wet thumb becomes cold in spite of the fact that the water from your mouth was warm.
This is because the water is evaporating from your thumb into the air. Evaporation uses energy and this is obtained by cooling your thumb.

The wet and dry thermometer works in a similar way. The wet bulb is cooled by the evaporation of water. When the air is dry evaporation is rapid; it cools the wet bulb more than when the air is humid and evaporation is slow. The difference in the readings on the two thermometers will vary with the humidity. The readings are taken and the humidity is then read from a chart.

To measure the *relative humidity* using a wet

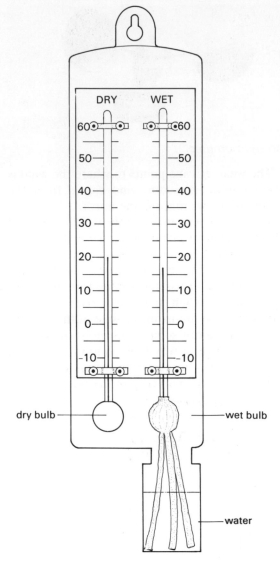

A wet and dry bulb thermometer

Relative humidity for use with wet and dry thermometer

Depression of wet bulb	Dry bulb temperature (°C)															
	0	2	4	6	8	10	12	14	16	18	20	22	24	26	28	30
1	81	84	85	86	87	88	89	90	90	91	91	92	92	92	93	93
2	64	68	71	73	75	77	78	79	81	82	83	83	84	85	85	86
3	46	52	57	60	63	66	68	70	71	73	74	76	77	78	78	79
4	29	37	43	48	51	55	58	60	63	65	66	68	69	71	72	73
5	13	22	29	35	40	44	48	51	54	57	59	61	62	64	65	67
6		7	16	24	29	34	39	42	46	49	51	54	56	58	59	61

and dry bulb thermometer, take the following readings:

1. The temperature of the dry bulb thermometer.
2. The difference in temperature between the two thermometers.
3. Put a ruler across the chart just under this difference.
4. Find the column with the dry bulb reading on top.
5. The figure you see at the bottom of this column is the relative humidity.

What is the relative humidity when the dry bulb reads 18 °C and the wet bulb reads 14 °C?

Practical 17.3

Use a wet and dry bulb thermometer to find the difference in humidity inside and outside your greenhouse.

Summary

1. The highest and lowest daily temperatures can be recorded with a *maximum and minimum thermometer*.
2. A *Stevenson's screen* is used to protect recording thermometers from sun and rain which would affect the readings.
3. The temperature of the soil varies with depth.
4. Certain trees, shrubs and plants must not be planted in *frost pockets* or *frost hollows*.
5. It is possible to record the hours in the day during which the sun shines with a *sunshine recorder*.
6. Wind speed is measured with an *anemometer*.
7. Wind speed can be estimated by using the *Beaufort scale*.
8. The wind which most often blows is known as the *prevailing wind*.
9. Wind direction is indicated with a *weather vane*.
10. *Rainfall* is measured as a depth of millimetres.
11. *Humidity* is measured with a *wet and dry bulb thermometer*.

Questions

1. Write single sentences to answer the following questions:
   (a) Why are less fields cropped with grass in the east than in the west of the country?
   (b) Which scale is used to estimate wind speed?
   (c) At what time of day, during a hot August, will soil at a depth of 20 cm be warmest?
   (d) Why is it necessary to avoid frost hollows when planting hardy fruit trees?
   (e) Why will a tree in a tub die in winter whilst a similar tree in the soil lives?
   (f) Why don't the iron riders in a maximum and minimum thermometer rust away?

2. (a) Name the instruments you would require to set up a weather station.
   (b) What recordings would you take, and when would you take them, if you were looking after a weather station?
   (c) Design a card suitable for taking with you to record your daily readings. The card should last for 1 week, when it would be replaced by a new one.

3. Which of the instruments described in this unit would be:
   (a) useful in a greenhouse;
   (b) of no use in a greenhouse?
   Give reasons for each of your statements.

Unit 18
Hen biology

The hen is an important link in our food chain. Hens are kept in nearly every country in the world, supplying protein and other important items to our diet.

Before these foods can be converted into eggs and meat the hen first has to break them down. This is done in two ways:

1. *Physical* – applying force to break the food into smaller pieces.
2. *Chemical* – applying substances to change the large molecules within the food into smaller ones.

The molecules must be small enough to pass through the wall of the gut into the bloodstream. The blood then carries the molecules to wherever they are needed in the body.

Light Sussex hen

Hen food chain

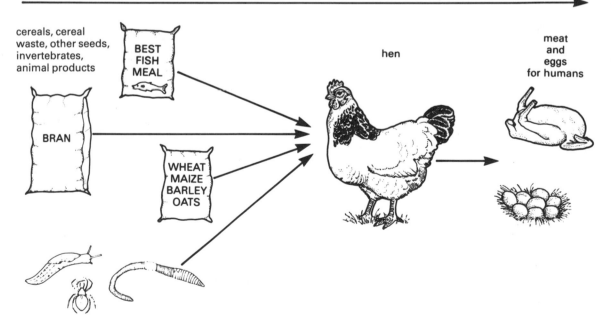

ENERGY

cereals, cereal waste, other seeds, invertebrates, animal products

BEST FISH MEAL

BRAN

WHEAT MAIZE BARLEY OATS

hen

meat and eggs for humans

The digestion process

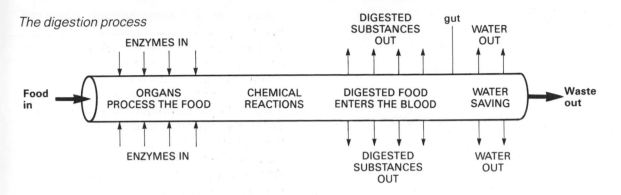

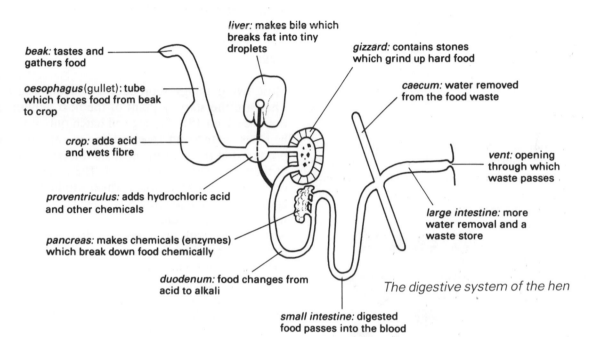

liver: makes bile which breaks fat into tiny droplets

beak: tastes and gathers food

gizzard: contains stones which grind up hard food

oesophagus(gullet): tube which forces food from beak to crop

caecum: water removed from the food waste

crop: adds acid and wets fibre

vent: opening through which waste passes

proventriculus: adds hydrochloric acid and other chemicals

large intestine: more water removal and a waste store

pancreas: makes chemicals (enzymes) which break down food chemically

duodenum: food changes from acid to alkali

The digestive system of the hen

small intestine: digested food passes into the blood

Digestive system

The breaking down of food into small molecules is known as *digestion*. This is done inside the bird by the *digestive system*. This system consists of a single tube (the *gut*) which runs through the body from the *beak* to the *vent*. As food is pushed along this tube it is processed by the various organs.

Task 18.1

1. Trace the diagrams on page 100 on to a piece of coloured paper.
2. Carefully cut out each piece.
3. Take a large piece of plain paper. Arrange the pieces to make up the complete digestive system of the hen.
4. When the pieces are correctly placed, stick them down.
5. Label each part.
6. List the parts. Say in your own words how each one aids digestion.

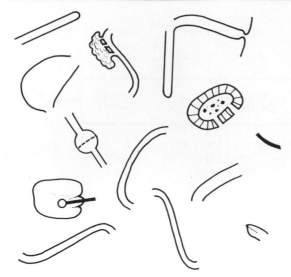

Diagram for Task 18.1

Gizzard

Note that the *gizzard* contains stones. Birds which are housed indoors should have access to small stones (these may be purchased as 'flint grit'). They will eat these in order to charge and recharge their gizzards.

Gizzard

The next experiment tests the idea that the hen can taste food with her beak.

Experiment 18.1

Items required: layer's meal; flavourings; a hen which is used to being handled; large cardboard box; garden canes; small pots

1. Obtain a range of food flavourings. You can increase the number by making some of your own (e.g. crush some garlic, onion, radish – or boil some strongly flavoured herbs e.g. rue).
2. Take about 50 g of layer's mash. Add a small amount of flavoured water. Mix until a crumbly texture is obtained.
3. Repeat using the other flavours.
4. Take a large cardboard box. Make bars down the front from garden canes set about 6 cm apart.
5. Place the hen in the box.
6. Offer it the pots of flavoured mash.
7. Observe and record.
8. Change the order of the pots and repeat.

Eggs

Most living things reproduce by laying eggs. They are the starting point of the next generation (see Unit 19). An egg will hatch only if the hen has mated and received sperm. *Sperm* is produced in the *testis* of the male; there are two of these situated inside the bird. The female has only one *ovary* (the organ which produces the egg) and one *oviduct* (the tube which leads from the ovary to the vent).

Note: Hens will lay eggs whether they are fertilised or not. It is not necessary to keep a cockerel with laying hens unless the eggs are needed for hatching.

If a laying bird is dissected the eggs which are developing in the ovary look like tiny egg yolks. These grow as they ripen and a single large yolk will then become detached and begin to move away. The free ovary then enters the oviduct; as it passes along *albumen* (egg white) and a *membrane* are added. The egg, now full size, passes through the shell gland and is coated with a layer of calcium carbonate – the *egg shell*.

The hen now has the complete, hard egg inside her oviduct. She seeks out a quiet, darkened place and sits until the egg reaches the vent. Then she stands up with her vent near to the floor and lays the egg blunt end first.

Notes:
1. The shell gland can produce only enough calcium carbonate to make one shell a day. If a hen lays two eggs in one day the second egg

Male reproductive system

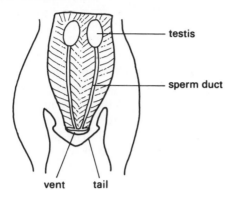

testis

sperm duct

vent   tail

Female reproductive system

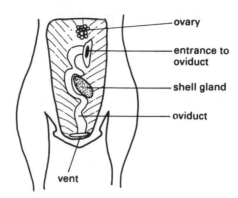

ovary

entrance to oviduct

shell gland

oviduct

vent

will be without a shell ('*soft shelled*').

2. Modern foods for laying birds contain enough calcium for the hen to convert into egg shells. If your school hens are eating green food and mixed corn they will require extra calcium when in lay. This can be offered as ground limestone rock or crushed sea shells. The hens will take just the amount they need.

Summary

1. The hen is an important part of our food chain.
2. *Digestion* converts foods into substances with smaller molecules.
3. The products of digestion pass into the blood.
4. Hens need a supply of small stones for their *gizzard*.
5. Digestion takes place inside the *gut*.
6. Hens and cockerels have their sexual organs inside the body.
7. Cockerels are not necessary for egg production, but only for *fertilisation*.
8. As the egg passes down the *oviduct* the *albumen*, the *membrane* and the *shell* are added.
9. Laying hens require *calcium* in their diets for shell formation.

Questions

1. Using words and diagrams, explain why laying hens should have access to:
   (a) flint grit;
   (b) crushed oyster shells.

2. (a) Explain what is meant by digestion.
   (b) Draw a diagram of the hen's digestive system.
   (c) Add these labels:
   crop, gizzard, duodenum, small intes-tine, vent, liver, caecum, pancreas, large intestine, proventriculus
   (d) Describe the function of:
   (i) the crop;
   (ii) the small intestine.

3. Using words and diagrams, explain why the second egg laid by a hen in a single day is usually soft shelled.

Unit 19
Eggs and incubation

An egg contains food, water and a female sex cell. If mating took place before the egg was laid the sex cell will be an *embryo* – a tiny speck which will grow into a young animal. As it grows the embryo uses up the contents of the egg. Almost all living things produce eggs. Some fish produce millions, laying them in the water and leaving them to chance. Most mammals keep their eggs inside their bodies. Here the eggs are safe and warm whilst they develop into young.

Birds' eggs are good food for other animals, which find and eat them. People eat the eggs of several different birds, including sea gulls, ducks, guinea fowl and quail. The hen's egg is our favourite; we have bred this bird to lay over 200 large eggs each year which we eat in many different forms.

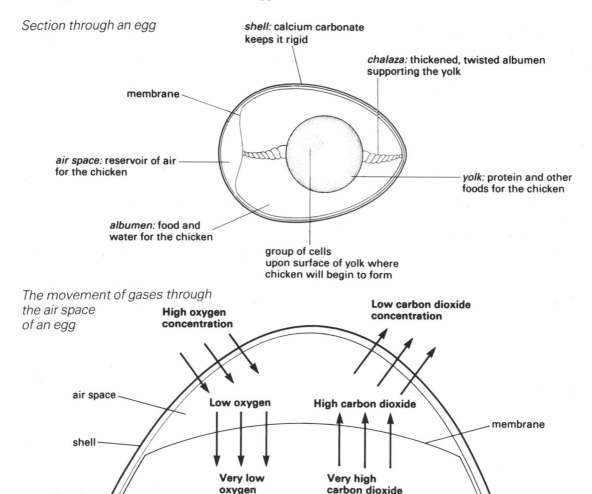

Section through an egg

shell: calcium carbonate keeps it rigid

chalaza: thickened, twisted albumen supporting the yolk

membrane

air space: reservoir of air for the chicken

yolk: protein and other foods for the chicken

albumen: food and water for the chicken

group of cells upon surface of yolk where chicken will begin to form

The movement of gases through the air space of an egg

High oxygen concentration

Low carbon dioxide concentration

air space

shell

Low oxygen

High carbon dioxide

membrane

Very low oxygen

Very high carbon dioxide

developing chick

The egg shell

The *egg shell* is a boundary that gases can pass through. Gases will move from an area of high concentration to one of low concentration. In the air space there is a high concentration of carbon dioxide and a low concentration of oxygen. So oxygen will move across the shell into the air space; carbon dioxide will move out.

The next experiment tests to see if the egg shell also slows down the loss of water from the egg.

Experiment 19.1

Items required: two eggs; dilute hydrochloric acid; beakers; test-tube brush; balance

1. Find the mass of one egg with an electronic balance.
2. Place the other egg in a beaker of dilute hydrochloric acid. Bubbles of carbon dioxide will appear as the acid reacts with the shell.
3. Gently brush away the bubbles with a test-tube brush and continue until all of the shell has dissolved away.
4. Wash away the surplus acid and dry the egg by gently rolling it on a paper towel.
5. Weigh the shell-less egg and record its mass.
6. Leave both eggs exposed to the air.
7. Weigh each egg daily for a week or so and record.
8. Plot the results as two *line* graphs on the same axes.
9. Record your conclusions.

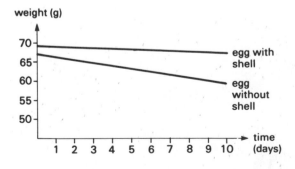

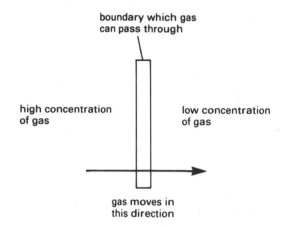

Gases passing through a boundary

Scientific note:
(a) We have assumed that any loss in mass is due to evaporation of water – this may or may not be true.
(b) Two eggs are not enough; the experiment would have to be repeated many times before any scientific conclusions can be made.

The shell also protects the egg from physical pressure. Try to break an egg by applying steady pressure to the ends. It is unlikely to break. Fortunately for the chicken, it is easier to break from the inside – why?

Trying to crush an egg

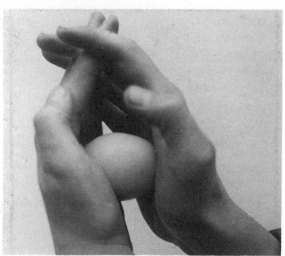

Egg grades

Eggs are graded by mass:

grade	1	2	3	4	5	6	7
mass (g)	>70	65/70	60/65	55/60	50/55	45/50	<45

Incubation

A *clutch* of eggs (the number a hen can cover and keep warm – usually about 12) may be incubated (i.e. kept warm until they hatch) by a *broody* hen. As a laying hen becomes broody her behaviour changes in the following ways:

1. She stops laying.
2. She spends a lot of time on the nest.
3. Her normal cackle changes to a cluck.
4. She ruffles her feathers.
5. When approached she does not go away, but sits tight.
6. She pecks hard at a hand which is put towards her.

If a broody hen is given her own box away from the other birds and provided with a dozen fresh, fertile eggs she will sit on them until they hatch. During the 3 weeks it takes for the eggs to incubate she will leave the nest only once each day to feed, drink and empty her bowels.

A broody hen:

1. Keeps the eggs warm and humid.
2. Protects them from predators such as crow and magpie.
3. Turns the eggs regularly.
4. Warms, feeds and protects the chicks when they hatch.

She cannot protect the chicks from foxes or rats, however, and her pen must be fox and rat proof.

Artificial incubation

Broody hens are often unreliable and not always available. It is usually better to use a small incubator. The one illustrated is made from expanded polystyrene.

An artificial incubator

Section through an incubator

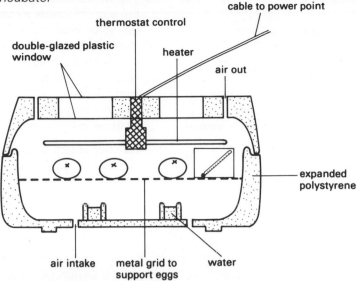

This is how to incubate *fresh, fertile* eggs using a polystyrene incubator:

1. Make sure that the incubator is thoroughly clean (do not scrub it).
2. Stand it on a hard flat surface. If stood on a cloth the air circulation may be affected.
3. Fill one ring with water.
4. Place the thermometer inside and replace the lid.
5. Plug in to a mains electricity supply and switch on.
6. Leave alone for 3 hours.
7. Read the thermometer; if the temperature is below or above 38.5 °C adjust the thermostat.
8. When the incubator has maintained a steady temperature of 38.5 °C for 3 hours or more, put a cross on to one side of each egg and put them into the incubator, all cross side up.
9. It will take some time to warm the eggs to incubation temperature. Leave it alone.
10. Turn the eggs twice a day; at the same time check that the water is filled up.
11. After 19 or 20 days there should be some sign of hatching – a 'pipped' egg or the sound of a chick cheeping from inside an egg.
12. Do *not* open the incubator, and do *not* assist any chicks; just leave the incubator closed for the next 24 hours.
13. Remove any fluffy chicks.
14. Give any remaining eggs another day to hatch.
15. Burn egg shells and debris, clean the incubator and store.

Practical 19.1 Candling eggs
(*1 week after incubation begins*)

Items required: wooden box; light fitting

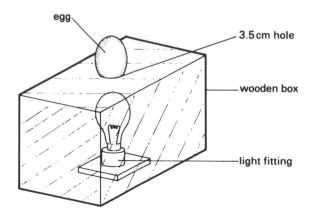

1. Drill a 3.5 cm circular hole in the bottom of a wooden box.
2. Invert the box over an electric lamp.
3. Dim out the room and switch off any lights.
4. Put a fresh egg, pointed end down, into the hole.
5. Observe the egg. Make sure that you can recognise the yolk and the air space.
6 Gently crack the egg shell and have a second look.
7 Remove the eggs from the incubator and examine each one in turn.
8. You will be able to tell the difference between the eggs in which a chick is developing and the 'clear' ones.
9. Note the air sac. Is it larger or smaller than the one in the fresh egg?
10. Return the fertile eggs to the incubator. Discard the clear ones.
11. Use the formula below to calculate the percentage of fresh eggs.

$$\frac{\text{number of fertile eggs}}{\text{number of eggs tested}} \times \frac{100}{1} = \%\ \text{fertile}$$

Summary

1. The *air space* in an egg acts as an air reservoir for the developing chick.
2. An *egg shell* is made of *calcium carbonate* (chalk).
3. An egg is a very strong structure which is difficult to break by applying a steady pressure.
4. Shop eggs are graded by *mass*.
5. A hen will only incubate eggs if she is *broody*.
6. A broody hen has a different behaviour pattern to one which is not broody.
7. An *incubator* keeps eggs at a constant temperature and humidity.
8. Incubating eggs have to be turned over twice daily.
9. It is possible to tell if an egg has a chick inside by *candling* the egg.

Questions

1. Describe each of the following parts of a hen's egg and explain their function (purpose):
   (a) shell; (b) air space; (c) chalaza; (d) yolk; (e) albumen.

2. (a) Draw a vertical section through the middle of an incubator.
   (b) Label the parts.
   (c) Draw arrows to indicate the flow of air whilst the incubator is in use.
   (d) Give two reasons why a supply of fresh air is necessary.
   (e) What would probably happen if you allowed the water tray to dry up?

3. Give a step-by-step account of how you would incubate 25 fresh, fertile hens' eggs using a small electric incubator.

4. Three hundred eggs were removed from an incubator and candled. Six of the eggs were found to be 'clear'. What percentage of the eggs were fertile?

Unit 20
Chickens

When a chicken hatches from an egg it is wet and should be left untouched in the incubator to dry. As it dries off the down covering on its body fluffs up and the chicken becomes active. (There is food and water within its body from the egg and it will survive without either for at least 24 hours.) It is during this period that chickens are transported from the hatchery to the farms where they will be reared. A small 'tooth' is present at the upper tip of the beak. This is an *egg tooth*; it was used to chip through the shell. It will drop off in a day or two.

Chicks in an incubator

The egg tooth

egg tooth

Baby chicks are unable to maintain their correct body temperature for two reasons:

1. They have a large surface area over which to lose heat and only a small mass to produce heat.
2. They have no feathers to keep their bodies warm and dry.

Chicks must be kept in a warm place until they and their feathers grow. This takes approximately 6 weeks. The apparatus which contains the chicks and provides the warmth is called a *brooder* and the artificial rearing of chickens is

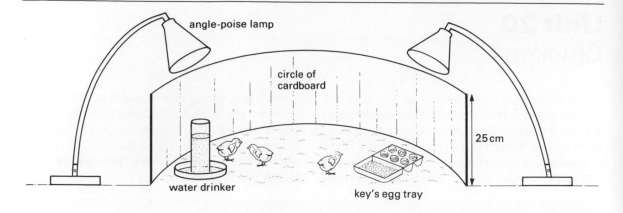

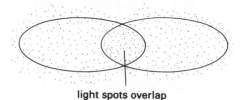

light spots overlap

known as *brooding*. There are many different types of brooder, some of which are very expensive. A few chickens (up to about 20) can be brooded by the following method:

The shed you use must be vermin proof as rats will take and eat young chickens.

1. Make a circle of cardboard about 1.5 m in diameter and 25 cm high.
2. Cover the floor with a layer of sawdust or peat (straw is not suitable unless it is chopped up).
3. Provide a spill-proof water container, filled with clean water.
4. Put a key's egg tray on the floor and sprinkle liberally with baby chick crumbs. This will be replaced after a few days with a food trough.
5. Fix two Anglepoise lamps so that the light-spots they make overlap:
6. Take the baby chicks from the incubator and place them under the lamps.
7. Observe the chickens and raise or lower the lamps until the chicks settle under them with the hottest spot left free. (If the lamps are too high the chickens will cluster into the warmest place.)
8. Food in the form of baby chick crumbs must be on offer all the time. This is known as *ab lib* feeding. Poultry which are fed ad lib do not overeat; they just take enough food each day to provide the energy they need.
9. Raise the lamps each week until the chicks are fully feathered and can maintain their

own body temperature.
10. It should now be possible to *sex* the chickens, that is to separate the *cockerels* (males) from the *pullets* (females) The rest of this unit refers only to pullets.

Note: With some breeds the males are a different colour to the females so the chicks can be sexed as soon as they hatch. One example is Rhode Island Red males crossed with Light Sussex females; male chicks are white and the female chicks are brown. Breeds with this characteristic are said to be *sex linked*.

Practical 20.1 Bird communication

1. When the chickens are about 3 days old, remove one and put it into an open-topped box a long way from the others.
2. After a short time the isolated chick will begin to make a distinctive noise.
3. Listen for a short while and then return the chick to the brooder.

The noise the chick made was the 'I'm lost' alarm call. If the chicken was being reared by a broody hen the mother would have responded to the noise by going to the chicken's aid.

Task 20.1

(*if you are rearing your own chickens weigh them weekly and use your own figures instead of these*)

The figures below show the average mass of 25 pullets reared from day old to point of lay.

Age (days)	3	10	17	24	31	38	45	52	59	66	73
Mass (g)	70	110	200	300	420	530	700	900	1050	1200	1350

Age (days)	80	87	94	101	108	115	122	129	136	143	150
Mass (g)	1450	1640	1755	2000	2120	2250	2370	2430	2480	2500	2500

Plot the figures on a *line* graph. Put age along the bottom and mass up the side.

Study the graph you have drawn and imagine that it is a hill up which you have to cycle. The places which are hardest to pedal up are the places where the chickens are growing most rapidly. Growth starts off slowly, then it quickens and finally at the top, when the bird is fully grown, it stops. (It is as though the top of the hill had been reached.)

Rearing chickens

After the brooding period chickens must be *reared* for about 12 weeks until they are mature enough to lay. This can be done extensively or intensively.

Extensive methods

The chicks are housed at night and allowed outside to roam freely during the day. Each bird requires a lot of space.

Intensive methods

The chicks are confined to a shed day and night. The floor is covered with a deep layer of straw or peat, which absorbs the droppings and allows the birds to dust bathe and to scratch. Each bird requires a small amount of space.

Whichever method is used, the birds need to have grower's mash or pellets on offer all the time and a continuous supply of clean water.

Each method has advantages and disadvantages. A third method which has some of the advantages of both methods and less of the disadvantages is the *mobile run* or *ark*.

The ark is moved regularly, giving the birds access to clean fresh grass. The floor of the pen consists of wooden slats through which the droppings fall. This removes the need for cleaning out.

A poultry ark

Task 20.2

1. Study the chart below of advantages and disadvantages of the two rearing systems.

2. Make a similar chart showing the advantages and disadvantages of a poultry ark.

	Advantages	*Disadvantages*
Intensive	Small amount of work Once a day attention Birds eat less food Safe from predators Lighting control is possible	Lot of litter needed Birds less hardy Vices (e.g. feather pecking) are more likely
Extensive	Birds are more hardy Vices like feather pecking unlikely Birds eat some natural food	Birds eat more food Predators may take birds Twice daily attention is needed Large area of land required

Whichever system is used for rearing, changes will occur in the birds as they reach maturity. The body shape changes as the hind parts enlarge. The comb and wattles grow rapidly and turn a brighter red. The birds are then on the point of *lay*; they can be moved to laying quarters, or left in the same house. Their food should be gradually changed from grower's mash or pellets to layer's mash or pellets. Darkened nest boxes with clean hay or shavings must be provided before the first eggs are laid. The chickens are now *hens*, although they may be called 'pullets' in some areas. The first egg a hen lays often has a small smear of blood on the shell. Do not be alarmed about this; it is quite normal. Hens will probably lay around 200 eggs each at the rate of five or six a week. The figure of 200 is an average. The actual number of eggs laid will depend upon the breed of the hens and the time of year when they began to lay. After the laying period the birds will *moult* (change their feathers) and have 2 or 3 months' rest from laying.

A dust balts: hens love dust baths as these keep them clean and parasite-free

The next experiment tests the idea that hens eat more food in pellet form than in powdered form (meal).

Experiment 20.1

Items required: spring balance

1. Use a pen of hens which are in full lay.
2. Set up a spill-proof hopper as shown in the photograph.
3. Fill with pellets and note the mass.
4. Record the mass of pellets consumed during a week.
5. Empty the hopper and refill with similar food which is in the form of meal.
6. Record the mass eaten during the week.
7. Repeat for two more weeks.

	Number of birds	Mass eaten	Mass per bird
Week 1 – pellets			
Week 2 – meal			
Week 3 – pellets			
Week 4 – meal			

Average of weeks 1 and 3 (pellets) = _____
Average of weeks 2 and 4 (meal) = _____

Spring balance

Taste

It is sometimes claimed that eggs have a better flavour if they are laid by birds which run in the fields finding some of their own food – the system known as *free range*. It would be very difficult to prove or disprove this as the taste depends upon the opinion of the person who is tasting. It is not possible to measure or weigh 'taste' in a scientific way.

Rat- and bird-proof hopper: hens kept on free range can lose food to wild animals. This hopper drops food only when the hen presses the trigger

Practical 20.2 Taste

Items required: new laid egg; shop egg; beaker; egg spoon

1. Collect an egg which has been laid that day. Grade it by weighing. (See Unit 19.)
2. Obtain an egg of the same grade which has been purchased from a supermarket.
3. Put the eggs into boiling water and boil them for 4.5 minutes.
4. Open both eggs. Are they set equally hard?
5. Observe any differences in the colour of the yolks.
6. Compare the taste of the two eggs. Can you tell the difference? If so, which do you prefer?

Summary

1. Baby chicks need to be kept warm (35 °C).
2. A device for keeping baby chickens warm is called a *brooder*.
3. Chicks are *brooded* for 6 weeks and *reared* for 12 weeks.
4. Chickens kept in intensive conditions may begin to peck each other's feathers.
5. *Intensive farming* is where a large number of birds are kept on a small area.
6. When chickens are changed from one type of food to another the change should be made gradually over a number of days.
7. Chickens eat just enough food to provide the energy they need and do not overeat when fed *ad lib*.
8. *Ad lib* feeding means having food on offer all the time.

Questions

1. Describe how to brood 25 baby chicks from 1 day old to 6 weeks old. Include at least one diagram in your description.

2. (a) Describe an intensive and an extensive method of rearing chickens from 6 weeks old to the point of lay.
   (b) What are the advantages and disadvantages of the methods you have described?
   (c) What changes can be observed in the birds as they approach the point of lay?

3. (a) Draw a plan (view from the top) of a poultry ark.
   (b) State two reasons why an ark should be moved every few days.
   (c) A 20-bird ark costs £150 and a 500-bird deep-litter house costs £2500. Calculate the cost per bird of each method of housing.

4. Ten chickens reared from 1 day old to point of lay ate 15 kg of baby chick crumbs and 60 kg of grower's pellets. Baby chick crumbs cost 25p per kilo and grower's pellets cost 20p per kilo.
   (a) What was the total cost of food used?
   (b) What was the cost of food per bird?

Unit 21
Birds

The area where an animal lives and feeds is called a *habitat*. School grounds form all or part of the habitat for a number of wild birds.

The school garden can increase the number of habitats available. This in turn increases the number of different species which visit the school.

Habitats or part habitats in school

Playing field	Insects for dunnocks Worms for thrush and blackbird
Playground	Dropped sandwiches, etc. attract carrion crows, gulls, sparrows, starlings
Buildings	Nest sites for sparrows, starlings, blackbirds, robins, house martins
Ornamental trees	Food for tits, thrushes, waxwings

There are a number of ways to encourage birds. You can provide them with a habitat. For instance, farmers encourage pheasants on their land by leaving corners of their cereal fields uncut, as in the photograph. At school or in your garden, you could provide a *bird box*.

A bird box

You can also encourage birds by providing them with food, either on a bird table or by setting up a bird area, on a horizontal branch of a tree, for instance. You will need to provide both food and water. A very good food is wild berries (e.g. elderberries), which can be collected in the autumn and stored in a deep freezer.

A corner of a cereal field left uncut to encourage pleasants

Practical 21.1

Items required: book for the identification of common birds

1. Observe the birds which are using your school grounds as part of their habitat.
2. Use a reference book to identify each species.
3. Suggest a reason for the presence of each one.
4. Record it on a chart similar to the one below.

Date	Bird	Suggested reason for presence
Sept. 14	Swallow	Stop-over during migration
"	Missel thrush	Feeding on worms on hockey field
"	Pied wagtail	Bathing in a playground puddle
"	House sparrow	Feeding on crumbs
"	Blue tit	Feeding on insects in cherry tree
"	Carrion crow	Eating a sandwich
"	Robin	Using a tree as a song post
"	House martin	Summer resident – nest under eaves

Why not set up a wildlife area in your school grounds? It will attract both plenty of birds and other animals.

Practical 21.2
(*best done in winter*)

Items required: bird food, e.g. seeds, dried fruit, crumbs, nuts; 50 g lard

1. Collect a good range of dry bird food.
2. Chop or crumble the large items and thoroughly mix it with the seeds and dried fruit.
3. Put the food into a plastic cup.
4. Melt the lard and pour it over the mixture.
5. Leave it to set.
6. Peel the plastic cup away from the block of 'bird pudding'
7. Set up a bird area in a place where it can be observed without disturbing the birds.
8. Keep the tree regularly stocked with food. When feeding, remember to sprinkle some food on the floor for the ground feeders like the dunnock and the wren.
9. Check the water level each day.
10. Observe and record.

Nests

The next experiment investigates *nesting*. It tests the idea that birds have no colour preferences when they are choosing their nest materials.

Experiment 21.1

Items required: old knitted woollen garments; cardboard; plastic net

1. Design and make a 'preference box'. This should have several compartments for different colours of wool.
2. Select three woollen garments with contrasting colours. (The garments must *all* be wool and not other types of fibre. Otherwise the birds may be selecting between wool and nylon, and not between colours. A good experiment will only have one *variable*.)
3. Cut a square of material with sides of 10 cm. Pull apart the fibres.
4. Repeat this for the other colours.
5. Weigh each colour accurately on an electronic balance and place it in the box.
6. Fix the box in a tree.
7. A few days later retrieve the box and weigh the fibres.
8. Record the results from the whole class on a table like that below.

Colour	Mass at start	Mass at end	% taken by birds
Red			
Green			

Adaptation

During your study of birds note how suited they are to their way of life. This is most obvious in the beaks and feet.

As an example of *adaptation* to life style, we will look in more detail at two birds: the hen and the goose.

BEAKS

CURLEW
Beak picks small animals and insects from the ground and probes into mud

EAGLE
Strong hooked beak for tearing flesh

HAWFINCH
Powerful beak to open seeds

SPARROW
Short strong bill – little adaptation – many different types of food

TREE CREEPER
Fine curved beak for picking small insects from the cracks in tree bark

FEET

HAWK
Strong feet with long curved claws for catching and killing prey

WOODPECKER
Two toes point forward and two toes point backwards, giving a good grip on a vertical tree trunk

HERON
Wide spreading foot on a long leg – for standing in water

The hen and the goose

Feet

Look at the hen's foot. Note how well adapted it is for both perching and scratching. The toes are tipped with robust claws which will cut into the soil or grip the perch. Three toes point forward to give a three-pronged scratching tool, whilst the one which points backwards gives balance for standing and grip for perching. The whole limb is protected with strong overlapping scales.

Left: hen
Right: goose

The foot of the goose is quite different as it is adapted to a very different life. It has the same pattern as the hen: four toes of which three point forward and one backward. The three forward toes have a large web of skin between them; the fourth toe is small and hangs free. This foot would be no good for either scratching or perching. It is, however, excellent for swimming and walking over muddy areas.

A goose lives by grazing, especially in areas where it is likely to be muddy. It also spends a lot of time on water. The hen's ancestors would have lived by scratching the forest floor searching for seeds and insects.

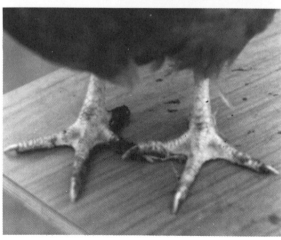

The feet of a hen

The feet of a goose

Heads

On the head of the chicken are large fleshy structures – the comb on the top and the wattles underneath. If a goose had a comb and wattles they would no doubt get frost bitten as the ancestor of the goose comes from much colder parts than the ancestor of the chicken. This is an example of how an animal is adapted to the environment in which it lives.

Look carefully at the two beaks. The goose's beak is much larger, stronger and less pointed than the chicken. A stronger beak is needed to graze grass, whereas a pointed beak is needed to pick small seeds from the soil.

The head of a hen

The head of a goose

The wing of a hen

Wings and feathers

The feathers of the two birds are also different. The goose, which spends some time on water, has closely packed, oily feathers. The goose also has a layer of *down* feathers next to its skin; these are such good insulators that they are used in the best quality continental quilts. It would not be possible for a goose to have a dust bath, which the hen often does. The hen gets grit right down to its skin and then shakes it off. As the grit flies off it takes specks of dead skin and other dirt with it.

In the wild the goose has to make long flights whilst the chicken only has to make short flights, between two lots of cover or up into a tree. Examine the wings – the goose wing is long,

The wing of a goose

pointed and adapted to open-air flying; the hen wing is short, rounded and much more suited to flights in a forest.

Practical 21.3 Comparison of the hen and goose

1. Make a large copy of the chart below.
2. Examine a hen and a goose. Look carefully at each part and complete the boxes.
3. Beginning with the feet, use the information above. Try to work out how each feature suits the bird's way of life.
4. If you do not have any birds available then work from the photographs.

Adaptation of the hen and goose to their way of life

Body part	Hen	Goose
Feet		
Legs		
Feathers		
Wings		
Comb and wattles		
Beak		

Summary

1. A *habitat* is an area where animals live and feed.
2. Several species of birds use school grounds as part of their habitat. It is possible to increase the number of bird habitats in a school.
3. Birds can be encouraged by providing *wild areas* or *bird boxes*, and by giving food and water.
4. Different birds are *adapted* to suit particular ways of life.

Questions

1. List the different species of birds which may visit your school. Suggest a reason for each visit.
2. Describe ways in which wild birds can be attracted to a school.
3. (a) Draw diagrams showing the foot of the hen and the goose. Explain how each foot is adapted to suit the bird's way of life.
   (b) Draw a diagram of the foot of a small wild bird which would be adapted to feeding on seeds gathered from thistles and similar plants.

Unit 22
Trees

A tree is a long-living woody *perennial*. There are two groups of trees: *broadleaves* and *conifers*.

Trees in a landscape: what would the countryside look like without them?

The broadleaf trees help make the countryside pleasant to look at, especially when they are growing in hedgerows and small woods. Nearly all the broadleaf trees are *deciduous* (shed their leaves in autumn); exceptions are holly, evergreen oak and box. The timber from a broadleaf tree is known as *hardwood*. It is used to make good quality furniture.

Conifer trees have leaves which are either *needles* or *scales*, and seeds are borne in *cones* (except yew and juniper). Almost all are *evergreen*; the exception is larch, which loses its needles in the autumn and produces new ones in spring.

Practical 22.1

1. Examine the twigs of a pine or a spruce.
2. Identify this year's needles. Find the junction on the twig which marks this year's growth.
3. Beyond this junction are last year's needles. Can you see any differences?

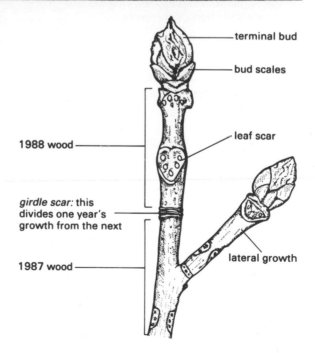

The external features of a twig

Labels: terminal bud; bud scales; leaf scar; lateral growth; 1988 wood; *girdle scar:* this divides one year's growth from the next; 1987 wood

4. Find the next junction on the twig and look at the older needles.
5. How old are the oldest needles?
6. Draw a diagram of the twig and label each group of needles with the year in which they first grew.

A conifer tree sheds its older needles a few at a time all the year round. Needles rot away very slowly; most forest floors are carpeted with them.

You can work out the age of a tree by counting the annual rings.

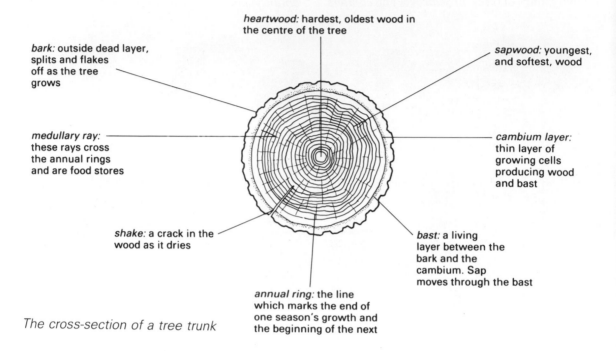

heartwood: hardest, oldest wood in the centre of the tree

bark: outside dead layer, splits and flakes off as the tree grows

sapwood: youngest, and softest, wood

medullary ray: these rays cross the annual rings and are food stores

cambium layer: thin layer of growing cells producing wood and bast

shake: a crack in the wood as it dries

bast: a living layer between the bark and the cambium. Sap moves through the bast

annual ring: the line which marks the end of one season's growth and the beginning of the next

The cross-section of a tree trunk

Task 22.1

Examine the cross-section of a real tree and answer the questions (assume that it has just been felled).

1. How old is the tree? When was it planted?
2. In which year was the worst growth (has the closest annual rings)?
3. In which year was the best growth (has the widest annual rings)?

Conifers produce *softwood*. This is the timber which most wood-using industries require. Conifer timber is also *pulped*; this wood pulp is used for making paper.

Most of our forests consist of conifers as they grow more quickly than broadleaves. Conifers also grow in areas where the soil and climate are not good enough for growing crops like grass, cereals or vegetables.

Ten per cent of the land in the United Kingdom is being used to grow trees. Ninety per cent of the timber used in the United Kingdom is imported.

Task 22.2

1. Make a list of the items which we make out of timber.
2. Make a second list of items we get from trees without cutting them down.

Trees have to be felled to provide timber. Do not forget, however, that trees are very important to humans whilst they are growing. They provide shelter and wind breaks, they remove carbon dioxide from the air and replace it with oxygen, and in many places they stop the soil from blowing or washing away and the snow from sliding down slopes. Trees provide food for us in the form of fruit and nuts, and natural rubber is made from the sap of a tree.

Trees also provide food and shelter for thousands of wildlife species. They make the countryside beautiful and also improve the towns.

Practical 22.2 Identifying trees
(*done during winter*)

Collect a number of leafless twigs from different trees and use the key in the next section to identify them.

A key for identifying trees from their twigs

1. Make sure that you understand the terms used (see the diagram of a twig earlier, page 119).
2. Look at the key. Each box has a number. Take a twig and begin at box number 1. The questions in the box refer to your twig. You will be able to answer *yes* to one of the questions in the box; if this question is followed by a number go down to the box with that number and carry on. Sooner or later when you answer 'yes' to a question you will be given the name of a tree. This is the *species* (type) of tree that your twig comes from.
3. Take a twig and begin at number 1. If the buds are *opposite* (see diagram) proceed to number 2; if the buds are *alternate* proceed to number 4, and so on.
4. Repeat this with other twigs.

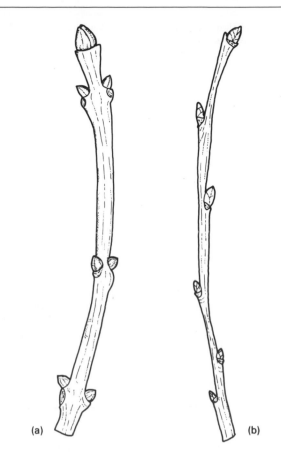

(a)

This twig has buds opposite

(b)

This twig has buds alternate

KEY

1. Are the buds opposite?	if *YES* go to 2
Are the buds alternate?	if *YES* go to 4

2. Are the bud scales black?	if *YES* it is **ASH**
Are the bud scales sticky?	if *YES* it is **HORSE CHESTNUT**
Are the bud scales slightly hairy?	if *YES* it is **MAPLE**
Are the bud scales untidy and is the twig smelly?	if *YES* it is **ELDER**
Are the bud scales green?	if *YES* go to 3

3. Are there two terminal buds and is the leaf scar narrower than the bud?	if *YES* it is **LILAC**
Is there one terminal bud and is the leaf scar wider than the bud?	if *YES* it is **SYCAMORE**

4. Are there prickles at the nodes?	if *YES* go to 5
Are there no prickles at the nodes?	if *YES* go to 7

5. Are there two prickles, one each side of the bud?	if *YES* it is **ROBINIA**
Are the prickles single?	if *YES* go to 6

6. Are the buds cone shaped?	if *YES* it is **WILD PEAR**
Do the buds have round tops with a prickle below the bud?	if *YES* it is **HAWTHORN**

7. Are there two bud scales and is each bud on a very short stem? — if *YES* it is **ALDER**

Are the buds without short stems? — If *YES* go to 8

8. Is the end of the twig hollow? — if *YES* it is **WALNUT**

Is the end of the twig solid? — if *YES* go to 9

9. Is the twig covered with lots of tiny scales? (with cones sometimes present) — if *YES* it is **LARCH**

Is the bark without scales? — if *YES* go to 10

10. Does the bud only have one scale? — if *YES* go to 11

Does the bud have more than one scale? — if *YES* go to 12

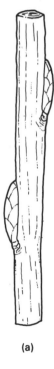

(a)

A bud pressing close to a twig

(b)

A bud sticking out from a twig

11. Are the buds pressing close to the twig (see diagram)? — if *YES* it is **WILLOW**

Are the buds sticking out from the twig and does the leaf scar go nearly all around the bud? — if *YES* it is **PLANE**

12. Are there a lot of buds right at the top of the twig? — If *YES* it is **OAK**

Are there a lot of buds at the top of short side shoots, and is the bark smooth? — if *YES* it is **CHERRY**

Are there just one or two terminal buds? — if *YES* go to 13

13. Are there just two bud scales on a lopsided bud? — if *YES* it is **LIME**

Do the buds have more than two scales? — if *YES* go to 14

14. Are the buds pressing close to the twigs? — if *YES* go to 15

Are the buds sticking out from the twigs? — if *YES* go to 18

15. Are the buds long, shiny and slightly sticky? — if *YES* it is **POPLAR**

Are the buds not sticky? — if *YES* go to 16

16. Is the stem woolly near to the terminal bud? — if *YES* it is **APPLE**

Does the stem appear smooth and not woolly? — if *YES* go to 17

17. Are the bud scales green with brown, hairy edges? — if *YES* it is **WHITEBEAM**

Are the buds long with brown scales with darker edges? — if *YES* it is **HORNBEAM**

Are the buds large, dark coloured and hairy? — if *YES* it is **PEAR**

Are the buds small, brown and cone shaped? — if *YES* it is **ROWAN**

18. Are the buds long, pointed and cigar-shaped? — if *YES* it is **BEECH**

Are the buds not shaped like long cigars? — if *YES* go to 19

(a)

Bud to one side of leaf scar

(b)

Bud over centre of leaf scar

19. Are the buds to one side of the leaf scars (see diagram) and are the twigs green without hairs? — if *YES* it is **SWEET CHESTNUT**

Are the buds to one side of the leaf scars and are the twigs brown and hairy? — if *YES* it is **WYCH ELM**

Are the buds over the centre of the leaf scars? — if *YES* go to 20

20. Are the buds silver-grey and hairy? — if *YES* it is **LABURNUM**

Are the buds smooth? — if *YES* go to 21

21. Are the buds small and arranged in a spiral? (with catkins often present) — if *YES* it is **BIRCH**

Are the buds on the lower part of the twig all rounded? — if *YES* go to 22

Are some buds pointed and some rounded? — if *YES* go to 23

22. Are the buds arranged in two lines on opposite sides of the twig? — if *YES* it is **HAZEL**

Are the buds arranged in a spiral? (thorns may be present) — if *YES* it is **HAWTHORN**

23. Are the twigs green down one side? — if *YES* it is **ALMOND**

Is the older bark wrinkled? — if *YES* it is **ELM**

Is the older bark smooth? — if *YES* it is **WYCH ELM**

Other ways of identifying trees

Tree shape

Horse chestnut

Fir

Oak

Poplar

Spruce

Leaves

Lime

Willow

Oak

Plane

Ash

Alder

Needles

Pine

Spruce

Larch

Cedar

Flowers

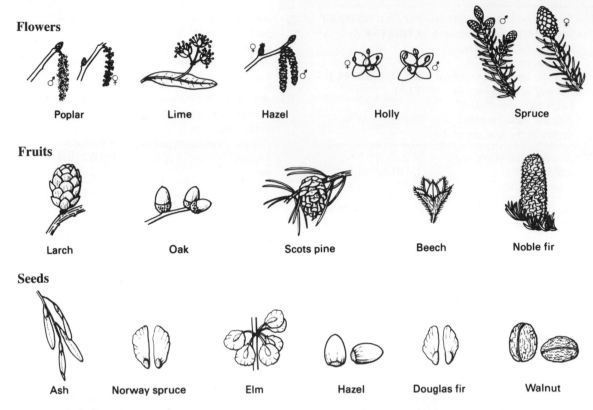

| Poplar | Lime | Hazel | Holly | Spruce |

Fruits

| Larch | Oak | Scots pine | Beech | Noble fir |

Seeds

| Ash | Norway spruce | Elm | Hazel | Douglas fir | Walnut |

Task 22.3: Leaf plaster casts

Task 22.3

Prepare a plaster cast of a leaf, like the ones in the photograph opposite, using the following method:

1. Take about 70 g of plasticine and soften it in your hands.
2. Press it into a flat smooth disc and cut off the edges to the required shape with a knife.
3. Press your leaf, underside down, into the plasticine to make an impression. Remove the leaf with care.
4. Take a strip of paper about 40 mm wide and fasten it around the plasticine with a little sticky tape.
5. Put about 20 mm of cold water into a disposable plastic cup.
6. Add plaster of Paris to the water (without stirring) until there are small dry islands upon the surface of the water.
7. Stir the plaster of Paris and pour it into the mould.
8. Leave for an hour to set.
9. Remove the paper and plasticine, and paint as required.

Summary

1. A tree is a *woody perennial*; it may be *broadleaf* or conifer, and *deciduous* or *evergreen*. Conifers have *cones*, and *needles* or *scales* for leaves.
2. *Hardwood* is timber from a broadleaf tree. *Softwood* is timber from a conifer tree.
3. The *girdle scar* on a twig separates one year's growth from the next. Other features are a *terminal bud*, *bud scales* and a *leaf scar*.
4. The cross-section of a tree shows annual *growth rings*. By counting these it is possible to tell the age of the tree. Other features of a tree trunk are: *sapwood*, *heartwood*, *bast*, *bark*, *shake* and *medullary rays*.
5. Living things can be identified with *keys*. To use a key, take a specimen and answer the questions in the first box; the answer will lead you to another box. Continue until the 'yes' answer gives the name of the specimen.
6. Twigs, tree shape, leaves, flowers, fruit and seeds all assist in the identification of trees.

Questions

1. Write single sentences to answer the following questions:
   (a) Give one use of hardwood.
   (b) What proportion of the timber that we use is grown in this country?
   (c) Which tree has bud scales which are green with brown, hairy edges?
   (d) How could an elm tree be distinguished from a wych elm?
   (e) In what ways does heartwood differ from sapwood?

2. (a) Make a list of all the items you can think of which are made from wood.
   (b) In what other ways are trees important to us?

3. Draw a cross-section through the main trunk of a 20-year-old tree. Include the following:
   (a) The tree had poor seasons during its 8th, 9th and 14th years.
   (b) The tree had a very good season during its 6th year and good seasons during its 16th and 17th years.
   (c) Two shakes.
   (d) Eight medullary rays.

Unit 23
Forest trees

There are a lot of privately owned forests in this country. However, the majority of forests are planted and managed by a government body – the *Forestry Commission.*

Seventy years ago, when the Forestry Commission began, its purpose was to produce as much *timber* as possible. It did this by planting large forests of conifers.

Nowadays, the Forestry Commission considers things other than timber production, for example *landscape* (the appearance of the countryside), *conservation* and *recreation.*

If a large area is planted with just one species (type) of tree:

1. It will appear drab and uninteresting.
2. If a disease enters the area it may spread very quickly.
3. If conditions favour a pest (e.g. pine looper moth or spruce aphid) the whole forest will be in danger.
4. There will be less wildlife.

Lines of young conifers in a forest nursery

Planting young forest trees

Most forests are planted with several different species of trees. Some trees are planted for reasons other than timber production. Hardwoods like birch are sometimes planted along the side of a forest road. There are two reasons for this:

1. To protect the forest trees which lie behind from fire.
2. To improve the landscape.

The leaves, flowers and fruit which grow on these birches will provide food for lots of creatures which would be unable to exist in an all-conifer forest.

Types of forest tree

Trees most often planted by the Forestry Commission for timber are as follows:

Scots pine (*Pinus sylvestris*)

This is the only *indigenous* (native) tree which is planted in our forests. Although it will grow almost anywhere in the British Isles, it is planted as a forest tree only in the Midlands and South East where the climate is warm and dry. The timber is known as *red deal*. The heartwood has a reddish colour. Red deal takes creosote readily. It is often used for railway sleepers and telegraph poles. It is also used for rafters and joists in houses and for garden sheds.

Corsican pine (*Pinus nigra*)

This is used as a forest tree as it grows fast and has a very straight trunk with only slender side branches. A native of the Mediterranean, it thrives only in the south and Midlands, where it is often planted. As it grows faster than Scots pine the annual rings of its timber are wider apart. It is of similar quality, however, and has similar uses.

Lodgepole pine (*Pinus contorta*)

This tree was used by the Indians for their lodges

or wigwams. It is planted as a forest tree because it grows quickly and survives in wet, cold areas where the Corsican pine would fail. It is one of the few trees which thrives on poor, peaty moorland soils. Its timber is very similar to Scots pine in both appearance and use.

European larch (*Latrix decidua*)

In winter this tree is without needles. The larch grows well in most parts of the British Isles except on the very poorest soils and where the winds are salty. The timber is the strongest of the softwoods. It is used for boat building, gates and fences. Each larch tree needs ample space and light. So they must be planted further apart than other conifers. Hence the timber production from a forest of larches is less than it would be with some other species. The larch most often planted is the hybrid larch (*Latrix eurolepis*) which is a cross-breed between the European larch and Japanese larch. This tree is larger and grows more rapidly than either of its parents. This is an example of *hybrid vigour*.

Norway spruce (*Picea abies*)

This is the tree most often used as a Christmas tree. Many thousands are grown for this purpose. The Norway spruce grows over large areas of northern Europe and was introduced to this country many years ago. It grows well in shallow soils and is planted in all but the coldest and wettest areas. The timber is known as *whitewood*; it is even in texture and holds nails well. It is strong and light. Norway spruce timber is used for house building, ladders, boxes and general joinery. It rots quickly when outside and must be treated with a wood preservative.

Sitka spruce (*Picea sitchensis*)

This tree is a native of Alaska, where it is exported from the port of Sitka – hence its name. It grows on very poor soils and tolerates high rainfall. It grows upright even in areas where the winds are strong and salt laden. So it is often planted in coastal regions. Sitka spruce requires at least 120 cm of rain each year. So it is not planted in the Midlands or the East. The Forestry Commission have planted more Sitka spruce than any other kind of tree. This is because it grows very rapidly and its timber is in great demand for making chipboard, hardboard, packing cases, joists and rafters, cellophane, sticky tape, rayon fabrics and paper.

Tree felling

Tree poles loaded on a trailer

Douglas fir (*Pseudotsuga menziesii*)

These trees are planted on the better soils and grow very large. One specimen at Powis Castle in Wales is nearly 70 metres tall. Douglas fir requires better soils than Sitka spruce and is often used to replant felled woodland. When growing in good soil it grows almost as fast as the spruces. The timber is of very good quality and is used in construction work, for telegraph poles and in paper pulp.

Western red cedar (*Thuja plicata*)

Western red cedar will grow in almost any part of Britain. However, it is not planted on the poorest soils, nor in the worst climates, as it would make only very slow growth. It is usually started off under a 'nurse' crop of larch or birch as it grows faster in light shade than in full sunlight during its first few years. Cedar timber is very durable; it is often used for greenhouses, sheds and bungalows.

Western hemlock (*Tsuga heterophylla*)

This tree is a native of North America. It is being planted in increasing numbers in this country as it grows well in our climate. It requires better soils than spruces and will not tolerate such harsh weather conditions. New plantations are started off in between other trees. This is because young hemlocks benefit from some shade. The timber is used for building, box making and paper pulp.

Task 23.1

Transfer the information about our forest trees to a chart with these headings:

Tree Common name	Scientific name	Soil type	Climate	Timber uses	Notes

Key for forest conifers

Take a twig from a forest conifer and use the key in the same way as the winter twig key in Unit 22.

1. Are the needles in pairs? — If *YES* go to 2
   Are the needles not in pairs? — if *YES* go to 4

2. Are the needles over 6 cm long; does the terminal bud narrow to a sharp point? — if *YES* it is **CORSICAN PINE**
   Are the needles less than 4 cm long; are the terminal buds long and blunt? — if *YES* go to 3

3. Are the needles blue/green in colour (bark of trunk slight orange/red)? — if *YES* it is **SCOTS PINE**
   Are the needles mid-green in colour (bark of trunk dull brown/black)? — if *YES* it is **LODGEPOLE PINE**

4. Are the needles held singly away from the twig? — if *YES* go to 5
   Are the needles very small, flat and lying close to the twig giving a fern-like appearance? — if *YES* it is **WESTERN RED CEDAR**

5. Are all needles held singly? — if *YES* go to 6
   Are the needles on older wood held in clusters? — if *YES* go to 8

6. Are the needles on tiny pegs (which remain on the twig after the needle falls but come away when the needle is pulled off)? — if *YES* go to 7
   Are the needles not on tiny pegs? — if *YES* go to 9

7. Are the needles bluish and with their end in a sharp point? — if *YES* it is **SITKA SPRUCE**
   Are the needles mid green and pointed but not sharp? — if *YES* it is **NORWAY SPRUCE**

8. Are the needles true green, with their twigs light brown? — if *YES* it is **EUROPEAN LARCH**
   Are the needles bluish green, with their twigs rust red? — if *YES* it is **JAPANESE LARCH**

9. Are the needles all of similar length? If you pull the needle away, does it leave a smooth, round scar? — if *YES* it is **DOUGLAS FIR**
   Are the needles of different lengths, and crowded along the twig in a random way? — if *YES* it is **WESTERN HEMLOCK**

Practical 23.1

1. Make a large pair of tree callipers from three pieces of wood. There is no need to fix them together as they can easily be held in place by two people when measurements are being taken.

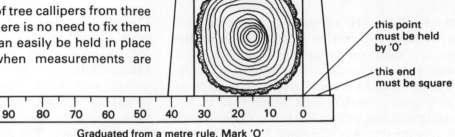

this point must be held by 'O'

this end must be square

90 80 70 60 50 40 30 20 10 0

Graduated from a metre rule. Mark 'O' 10 cm from the end.

Tree callipers

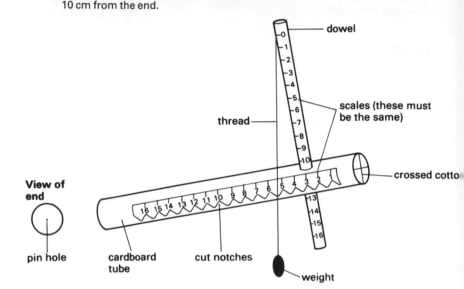

dowel

scales (these must be the same)

thread

crossed cotton

View of end

pin hole

cardboard tube

cut notches

weight

2. Make a hypsometer:

(a) Get a long cardboard tube.
(b) Stretch two threads across one end to cross in the middle. Secure it with tape.
(c) Cover the other end with card. Make a pin prick in the centre.
(d) Drill holes through the tube, towards one side. They should be the same diameter as a piece of wooden dowel. Make sure that you hold the drill at right angles to the tube.
(e) Mark a convenient scale on to a piece of dowel and the *same* scale on to a piece of stiff card. Put the dowel through the holes in the tube.
(f) Cut notches along the bottom of the card (dressmakers' pinking shears are ideal for this).
(g) Carefully glue the card to the tube.
(h) Fasten a thread to zero on the dowel. Use a small piece of plasticine to form a plumb bob.

You now have a *hypsometer*. It measures the height of trees.

How to use a Hypsometer

1. Measure a distance from a tree – say 20 metres.
2. Set the dowel rod at 20 (i.e. your distance from the tree) by sliding it up or down in the tube.
3. Keep the plumb bob hanging free. Sight the top of the tree through the pin hole and the crossed threads (like a soldier sights a rifle).
4. Keeping the tree top in sight, rotate the tube until the thread falls into a notch.
5. Take the reading on the scale along the tube.
6. Add the height of the observer's eye to get the height of the tree. This height will be in metres.

Practical 23.2

In order to do this work you need to visit a forest. However, parts of the work may be done in the school grounds using any trees which may be there.

1. Using the key on page 129, identify the conifers present in the forest. Record them in your note book.
2. Select an area of forest where the trees are planted regularly. Using a tape find out how many square metres of ground each tree occupies. For example, if the trees are planted 5 m apart in rows which are 6 m apart, each tree will have 6 m × 5 m = 30 m². To find the number of trees in 1 hectare use this equation:

$$\frac{\text{number of trees}}{\text{growing in 1 hectare}} = \frac{10\,000}{\text{area per tree}}$$

3. Measure the height of a number of trees. Calculate the average height.
4. Use the tree callipers you made to measure the diameter (at breast level) of a number of trees. Calculate the average diameter.
5. Using your results, calculate the amount in cubic metres of timber in an average tree using this formula:

$$\frac{\text{diameter in cm}}{2} \times \frac{\text{diameter in cm}}{2} \times \frac{3 \times \text{height in metres}}{5000}$$

This may look difficult, but it is easy with a calculator. You can now calculate the amount of timber in 1 hectare.

Note: One hectare = 10 000 square metres. This is the area of a square with a side of 100 metres.

Summary

1. The *Forestry Commission* is responsible for the majority of our forests.
2. Planting forests with a mixture of different tree species has a number of advantages over planting a single species.
3. Most forests are planted with *softwood* trees. Most of the timber used in this country is softwood.
4. Trees are planted for *timber, landscape, conservation* and *recreation*.
5. With the exception of Scots pine, all the forest conifers are species which have been introduced from abroad.
6. The height of a tree is measured with a *hypsometer*.
7. The volume of timber in a tree can be calculated from its height and girth measurements (measured by *tree callipers*).

Questions

1. Write single sentences to answer the following questions:
   (a) What is a hypsometer used for?
   (b) Which species of tree has orange/red bark?
   (c) Name the parents of the hybrid larch.
   (d) Which tree species was used by the Indians to construct their wigwams?
   (e) Which deciduous trees are conifers?

2. A Douglas fir tree was 45 metres tall and had a girth of 1 metre. What volume of timber would the forester have if this tree was felled?

3. Why is it better to plant several different species of tree in a forest rather than just a single species?

4. Ten hectares of forest are to be planted with Sitka spruce. The trees will be planted 5 metres apart along rows which are also 5 metres apart. How many trees will be needed?

Unit 24
The grey squirrel

The grey squirrel (*Sciurus carolinenis*) is a *rodent*. It was first introduced to the UK from North America during the nineteenth century. By 1915 they were numerous in many areas. Only the Lake District, East Anglia, the Isle of Wight, Anglesey and large areas of Scotland are free from this pest.

The grey squirrel is somewhat larger than the native red squirrel. It has no ear tufts and its tail is much less bushy; it is speckled grey with white underparts. The head and body measures 300 mm and the tail is 240 mm long.

The grey squirrel is an excellent climber and *arboreal* (tree living) in habit. It lives in forests, parks and open woodlands where there are deciduous trees, favouring areas containing beech. It does not live in areas where all trees are conifers; this confines it to the southern latitudes.

Food

Grey squirrels spend much time on the ground. They are *omnivores*, feeding on the buds and shoots of trees, fruit, puff-balls, toadstools, seeds, nuts, the seeds from pine cones, birds, birds' eggs and nestlings. The first item of diet mentioned ruins many trees, especially when the top buds are taken (as shown in the figure). In addition, during the months of April, May and June the bark is stripped from young beech and sycamore trees, as in the photo. This kills them. Why the grey squirrel strips bark is not known; it could be feeding on the sugars underneath the bark, or the action could be part of its social behaviour. During the late summer and autumn much time is spent burying food. Acorns, nuts and even fungus are buried under moss, or in small holes scratched in the soil. Food is buried over a large area, and is located during the winter by smell – the squirrel can detect it even when the ground is covered with snow.

The top of a young spruce tree: if eaten by a grey squirrel it would grow as in the diagram opposite

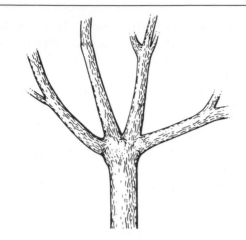

Tree growth after terminal bud destruction by a squirrel; a tree of this shape is of little value

A 25 year old sycamore tree that has been barked and killed by grey squirrels

Reproduction

The grey squirrel builds a nest of leafy twigs, lined with leaves, grass and bark – the bark of honeysuckle being favoured. The nest is called a drey. The drey is domed but has no fixed entrance, so the animal pushes in through the side it happens to be approaching from. The drey is usually situated against the trunk of the tree, supported by a branch. In the summer a more temporary drey is constructed out on the branches.

The young are born in the drey. There are two litters of three or four young each year, the first in the February–April period and the second during June–August. The young are born naked and blind, and they leave the nest after 8–10 weeks. They are fully grown at 8 months.

Control

Due to their arboreal habit, it is impossible to fence young trees against the grey squirrel. So the animals have to be controlled by the forester. His control methods include *shooting, trapping, drey poking* and *poisoning*.

Drey poking

Drey poking is possible only when the trees have no leaves. One man carries a gun and has a dog; another man has two 10 metre long aluminium poles which fit together. The upper end is hooked. The drey is touched lightly and the squirrel is shot as it puts out its head to investigate. The drey is then destroyed. This method is not very effective in clearing an area, as some squirrels could be in holes in trees and others away in search of food.

Poisoning

Poisoning with warfarin is practised in some forests. A special hopper is used to prevent other wild animals and birds from taking the poison.

Birds are not very resistant to warfarin, but they are unlikely to enter the hoppers and take the poisoned grain. Wood mice and voles can enter and are poisoned. These animals have only small territories compared with the grey squirrel. As only one hopper is placed in every 3 hectares,

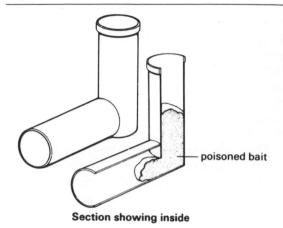

poisoned bait

Section showing inside

A poison hopper

only a small proportion of the small mammal population is killed; losses are soon made good from breeding in surrounding areas.

In order to prevent the *red squirrel* (Britain's native squirrel – a protected species) from taking poison, the law forbids the use of warfarin in areas where there are red squirrel populations (except inside buildings).

It is not known what effect warfarin has on the *secondary consumers* that feed upon mice and voles who are taking poison from the hoppers.

Trapping

Spring traps
Skilled *trappers* can find natural tunnels used by grey squirrels: hollow tree roots, dry drains, holes in banks, walls and hollow stumps. Where no natural tunnels are found they are constructed by driving stakes alongside a log. *Spring traps* are placed in these tunnels. They catch and kill any animal that steps upon the release plate. No baiting is required for this type of trap.

Cage traps
Cage traps can be used by unskilled people and are probably the most effective method of reducing the grey squirrel population. Trapping is carried out in the summer months when the squirrels are actively damaging the trees by stripping the bark from them. The forester considers trapping at other times of the year to be a waste of time, as the life cycle of this rodent

gives a high rate of replacement. This, together with quick dispersal, means that areas cleared in winter can be very badly damaged again the following summer.

Cage traps are baited with maize or acorns. When the animals enters to feed upon the bait the door closes behind and it is caught. The operator places a sack over the cage entrance and opens the door. The trap is then tilted with the door at the top and the animal instinctively runs upwards into the sack. It is then humanely killed.

NB *Shooting, trapping, drey poking and poisoning of squirrels or any other animal should only be practised by experienced and authorised persons.*

Summary

1. The grey squirrel is an *arboreal* (tree-living) *rodent*.
2. The grey squirrel was introduced from America. Our native squirrel is the *red squirrel*; it is a protected species.
3. Grey squirrels are *omnivores*.
4. Grey squirrels make a nest of twigs – these are called *dreys*.
5. Grey squirrels cause damage to trees, especially broad leaf trees.
6. Foresters attempt to control the numbers of grey squirrels by *shooting, trapping, drey poking* and *poisoning*. Traps include *spring traps* and *cage traps*.

Questions

1. Using words and diagrams, explain how the grey squirrel can cause damage in a forest.

2. (a) What methods do foresters use to control the numbers of grey squirrels?
   (b) You are told that only experienced and authorised persons should practise grey squirrel control. State *two* reasons why amateurs must not be allowed to control wild animals.

Unit 25
Wildlife

The term *wildlife* includes all plants and animals such as fungus, moss, worms, spiders and insects as well as flowering plants and vertebrates (animals with backbones like snakes, birds and badgers), living in their *natural habitat*.

Deciduous closed canopy

In a deciduous closed canopy, light reaches the ground after leaf fall. In early spring, this enables short-lived plants such as bluebells and anemonies to grow.

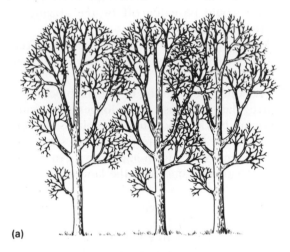

(a)

Evergreen closed canopy

In an evergreen closed canopy, light is excluded all the year round.

Trees

We saw in Unit 21 that a habitat is the area in which a creature lives or a plant grows. It provides food and shelter for animals and a place to grow for plants. A *mixed woodland* contains many different habitats; it is therefore home to many different plants and creatures. An area which has been close planted with *coniferous* trees is not a good place for wildlife because it provides a very small number of different habitats.

Tree habitats

Closed canopy
In a *closed canopy*, the tree tops touch and stop light from reaching the ground. There are few habitats for wildlife. There are two types: deciduous and evergreen:

Open canopy
In an *open canopy*, the tops of trees do not touch. Light falls on the ground all the year round. A field layer and a shrub layer form. This increases the number of habitats.

Open canopy

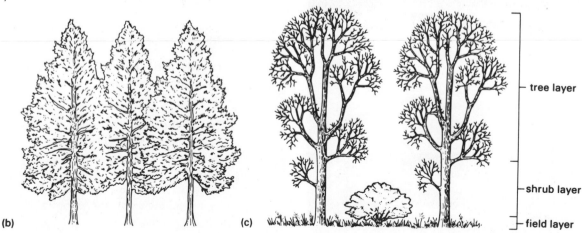

(b)

(c)

tree layer

shrub layer

field layer

The *tree layer* provides habitats for:	rook, squirrel, kestrel, pigeon, jay, missel thrush, tawny owl, tree creeper, woodpecker, starling, tits.
The *shrub layer* provides habitats for:	robin, wren, blackbird, gorse, bramble, azalea, broom, rhododendron.
The *field layer* provides habitats for:	woodmouse, fox, badger, mole, woodcock, vole, grasses, furze, ferns, bracken.

Dead trees provide food and shelter for many different creatures. If you can find a dead tree, peel off some bark and observe the creatures underneath. Many of our common birds need mature trees on which to build their nests. For instance, the owl needs a hollow tree in which to nest; it is unlikely to find one in a newly planted forest.

The Forestry Commission manages forests with the aim of producing as much timber as possible. They also, however, take steps to encourage and conserve wildlife. These steps include:

1. Allowing areas of open canopy.
2. When planting conifers in an area where there are a few mature broadleaf trees, leaving some of these broadleaf trees.
3. Planting broadleaf trees around the outside of plantations, around ponds and alongside rides.
4. Planting species which provide large numbers of habitats, e.g. oak, birch, willow, elm, Scots pine, aspen, alder, beech, hawthorn, blackthorn and hazel.
5. In forests which may be grazed, *fencing* off areas to allow plants to develop which otherwise would be eaten off by the grazing or browsing animal.
6. Making *footpaths* which keep people in certain areas and prevent trampling of delicate habitats.
7. Avoiding the planting of square blocks. An irregular block has a much longer edge than a square block. Woodland edge plants are very important for wildlife. These are similar to the plants which form hedgerows. An irregular shape also improves the landscape.

Coppice

When a coniferous tree is felled the remaining stump usually dies and decays. When a broadleaf tree is felled young shoots grow from

A fence protecting a young oak tree against browsing animals

The base of a coppice tree

around the remaining stump. If left, they grow thicker each year and soon become useful timber. Some trees (e.g. cherry) are grown especially to produce this kind of timber; the area in which these trees grow is known as a *coppice*.

Coppice trees are cut back to their stumps

Pine looper moth – a forest pest

The larva of the *pine looper moth* feeds upon the needles of pine trees. A small population of these will have no effect on a large pine tree. Occasionally, however, there is a *population explosion*, and the moths may eat all the pine

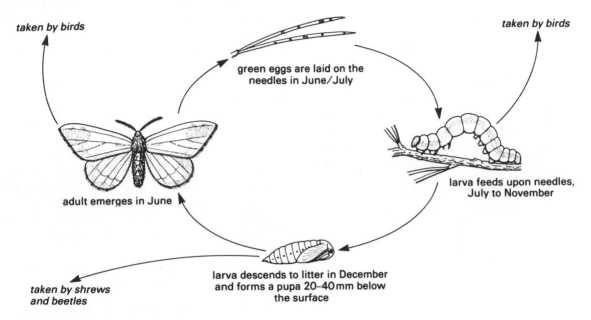

The life cycle of the pine looper moth

every 7 years or so. An area will be divided into seven parts and one part will be harvested each year. The timber from a coppice has different uses to timber from mature trees. It is used for fencing, rustic work, firewood, etc. A coppice provides a large number of habitats; it is therefore very rich in wildlife.

Task 25.1

Look at the information above about coppice trees. Make a list of all the reasons you can think of why a coppice is rich in wildlife.

Some forms of wildlife are good for the trees; others can be harmful; others do some good and some harm. A species causes real damage only when its numbers increase to a point where the trees begin to suffer.

needles. Then if the tree is not killed its growth will certainly be affected.

An all-pine forest with little wildlife is much more likely to suffer damage from the pine looper moth than is a mixed forest which is rich in wildlife.

Practical 25.1
(*This is best done during May about 4 weeks after the oak has come into leaf*)

Items required: plastic sheet at least 2 metres square; half a metre of a stout broom handle; collecting jars; pooter.

1. Find an oak tree with a low leafy branch. Four pupils need to hold the plastic sheet out on the floor underneath the branch.
2. A fifth pupil gives the branch a single hard bang with the broom handle.
3. Collect the animals which fall on to the

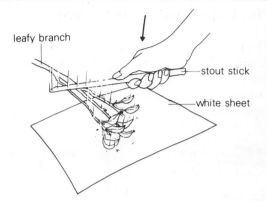

leafy branch

stout stick

white sheet

sheet, by hand or with a plastic teaspoon. Use a pooter where necessary. A pooter is a device for picking up small invertebrates; it can easily be made in the laboratory from a jar and glass tubing. Place the animals in jars, using different jars for different creatures.

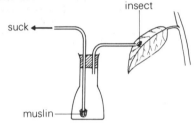

suck

insect

muslin

4. Count the animals in each jar.

Record your results on a bar chart or a pie diagram.

If this practical is done later in the year the 'catch' will be very much less. This is because oak leaves are only nutritious when young.

Task 25.2

A group of Staffordshire pupils had the following catch from a single oak bough in May:

mites	58	harvestmen	5
spiders	34	aphids	70
ladybirds	9	beetles	12
insect larvae	114	unidentified	58

1. What was the total number of invertebrates collected?
2. Display these results on a bar chart or a pie diagram.

Deer

The largest forest animal is the *deer*. It is also the largest wild animal in this country. Deer are very important; they attract many visitors to the forests and also provide meat known as *venison*.

Hundreds of years ago, deer herds were kept in a healthy condition by *wolves*, which hunted them for food. Wolves would catch the sick and the old animals. They would also cause large herds to break up into smaller groups. Now there are no natural predators, so deer wardens have to manage the herds and control their numbers. Too many deer in one forest would cause problems. They damage trees in several different ways:

1. They are browsers and eat the growing shoots of young trees. This causes the trunks to fork instead of growing into long, straight and useful timber (see Unit 24 pages 132–133)
2. They strip the bark from trees (especially in winter), which has the effect of killing the tree.
3. They rub their antlers against trees to remove the 'velvet'. This also causes damage.

Velvet is the skin and blood vessels which cover the growing *antlers*. The deer rubs it off when the antlers are fully grown. Deer shed and regrow their antlers each year. So this damage is an annual event.

A forest supports a certain number of deer without suffering any serious damage. The deer warden has to keep the deer population within that number. This is done by *culling*. This is removing the old and sick animals by shooting them carefully with a rifle so as to cause no pain. Some healthy animals may also have to be culled to keep the numbers down.

The size of the deer herd is regulated by their numbers in their three 'hungriest' months of the year: January, February and March. The number of deer which can be supported during these months without causing damage can easily be supported for the rest of the year.

In some forests deer are killed as they cross the road and so culling is not necessary. This is bad as the animals which are accidentally killed may not be the old and sick. They may be pregnant females. Worse still, they may be females suckling their young, and the fawns will die from starvation. As well as causing road accidents the deer may be injured and suffer a painful death. In an attempt to reduce the number of deer which are accidentally killed, *deer mirrors* are put near favourite crossing points. The deer are warned about approaching vehicles by the bright reflection of their headlamps in the mirrors. The clearing of roadside verges to improve visibility for both animals and drivers also helps. A surprising method is the removal of roadside fencing. If there is an average-height fence the deer will leap and so approach the road at speed. If there is no fence the deer can approach more cautiously. Where roads are fenced the fence has to be very high, as seen in the photograph.

There are four main species of deer which are present in the forests of the United Kingdom. These are:

Red deer (*Cervus elaphus*)

This is the largest species; it has a shoulder height of more than 1 metre. The *stags* (males) carry antlers which are replaced each year with a larger set. Antlers are *cast* in March and are fully grown for the *rutting* (mating) season the following October. For most of the year the stags and *hinds* (females) live in different areas. The hinds and their *calves* (young) have the best browsing areas. During October the males roar and attract the females for mating. They use their antlers to mark their territories and fight off

A walkers' gate over a deer fence

Forest deer will invade farm land and do great damage to crops. One method of reducing this without using expensive fencing is to create *deer lawns*. These are grassed areas in the forest that are cleared and sown especially for the deer to browse upon.

other males, although most of the 'fights' are settled without direct physical contact.

Calves are born singly the following May and suckled until January. The young stags stay with the hinds until they are two years old, when they grow their first set of antlers.

Red deer are found in the Scottish Highlands, the Lake District, Exmoor and some of the southern counties.

Roe deer (*Capreolus capreolus*)

The male is called a *buck*; it is 80 cm high at the shoulders. The antlers are much smaller than in the red deer; they are cast in December and fully grown by April. The roe deer does not live in herds but stays in family groups throughout the year. The female is called a *doe*; she usually gives birth to twins or triplets. Roe deer are common in Scotland and the Lake District and are very often seen in the home counties.

Muntjac (*Muntiacus reevesi*)

These small deer (shoulder height 55 cm) were imported from China over a hundred years ago by the Duke of Bedford for his parks on the Woburn estate. Several escaped and began breeding in the wild. They have now spread northwards from Bedfordshire and are present in many of the Midland forests. Unlike the native deer they do not have a rutting season. The does can have *fawns* at any time of year. The bucks have very small antlers, only 10 cm long. The muntjac has a 'hunch-backed' appearance and spends most of its time hidden in the undergrowth.

Fallow deer (*Dama dama*)

This is the most common deer species in the United Kingdom. Herds are found wild in most of Britain's larger forests. The fawns are spotted. They lie still in the undergrowth whilst the hinds browse.

The buck's shoulder height is 90 cm. Like other species it casts its antlers annually. The antlers are broad and flat; they are quite different to the round-sectioned antlers of the red deer.

People who walk their dogs in the forest during the breeding season should keep them under strict control when they are in deer areas. The dogs may find and disturb a young fawn.

shoulder height 55 cm

Task 25.3

1. Prepare a chart like the one below with as many horizontal columns as possible.
2. Enter information into each box.

	Red deer	Roe deer	Muntjac	Fallow deer
Shoulder height				
Name of male				
Name of female				
Antler type				

shoulder height 90 cm

A fallow deer

Summary

1. An area in which a creature lives is called its *habitat. Mixed woodlands* contain more habitats than *conifer forests*. Woodlands with *open canopies* have more wildlife than those with *closed canopies*. Open canopies contain a *tree-layer habitat*, a *shrub-layer habitat* and a *field-layer habitat*.
2. There are several ways in which the Forestry Commission can increase the numbers and variety of wildlife in the forests.
3. Coppice trees are cut every 5–7 years. A coppice is usually rich in wildlife.
4. Some wildlife can be harmful to trees.
5. If there is a wide variety of wildlife in a forest no single species is likely to become a *pest*. In all-conifer forests insects such as the *pine looper moth* can become pests.
6. A *pooter* is used to pick up small invertebrates.
7. There are four different species of deer common in the UK: the *red deer, fallow deer, roe deer* and *muntjac.*
8. Deer are an important asset to a forest. But their numbers have to be controlled to avoid too much tree damage. This is done by *culling. Fencing* and *deer lawns* are other ways of reducing damage.
9. Deer shed and regrow their *antlers* each year. Antlers are covered by *velvet.*

Questions

1. Explain why:
   (a) the close planting of large areas of pine trees can lead to a serious outbreak of pine looper moth;
   (b) an irregular-shaped area planted with forest trees is better for wildlife than a similar area planted as a square block;
   (c) making footpaths through a forest can aid wildlife.

2. Explain in detail why there are likely to be:
   (a) large populations of owls in an open canopy forest with an assortment of tree species of different ages;
   (b) small populations of owls in a closed canopy forest of a single tree species all of the same age.

3. (a) What is a coppice?
   (b) How can a coppice be organised so that a crop can be taken each year?
   (c) Examine the list of timber uses below and say which ones coppice timber would be (i) suitable for; (ii) unsuitable for:
   furniture; fencing; rafters and joists for house building; rustic work; pit props; paper making; fuel
   Note: Rustic wood is thin poles with the bark intact.

4. (a) Why does the deer warden have to control the deer numbers?
   (b) What steps can be taken to prevent:
   (i) Deer from being killed on the roads?
   (ii) Deer from invading farm land?
   (c) How did wolves keep deer herds healthy?

Unit 26
The badger and the fox

The badger

Badgers (*Meles meles*) are difficult to observe as they are *nocturnal* (active at night). You are much more likely to see one that has been killed on the road than a live one. They live in holes called *sets* in the ground. During the day they are deep inside these, warm and safe, curled up in a dry bed lined with leaves and grass.

Their sets are easily found as just outside the entrances are very large mounds of earth. There may also be tufts of old bedding material like straw. Scratch marks 1 metre high can often be found on nearby trees, and well-worn tracks often lead towards their main feeding grounds.

Badgers are present all over the British Isles except for heavily built-up and mountainous areas. They are most numerous in areas where there is a good mix of deciduous trees and farmland.

Food

It is not easy to find out what a shy nocturnal animal eats. It has taken a lot of research by

naturalists. They use the following methods:

1. Examination of the contents of the stomach of a dead animal.
2. Analysis of *droppings* – these consist of those parts of the food which have not been digested. Rabbits' fur, fruit seeds and beetles' wings may be recognised.
3. Observation of animals eating – this is difficult with a shy, nocturnal animal.
4. Observation of food preferences of animals in captivity.

We now know that the badger eats the following foods. First there are small young mammals, especially rabbits, whose nests are detected by smell and then dug out of the ground. Rats,

A badger set

mice, moles and hedgehogs are regularly taken; the hedgehog is skinned and the skin (prickles and all) is turned inside out and left. *Carrion* (dead flesh) is eaten, including dead lambs, but it is extremely unlikely that a badger would kill a lamb. Other food includes frogs (*not* toads), slugs, snails, beetle larvae, wasps and their nests, beetles and worms. The earthworm is probably the badger's most important food, especially on a damp night when they come to the surface. When eating a nest of wasps or wild bees, badgers become very excited and their hair stands on end; this probably prevents the insects from stinging them.

In addition, badgers eat a fair amount of plant material, including fruit (fallen apples, blackberries), seeds (acorns, beechmast), cereals, bulbs (particularly bluebells), roots and grass.

The badger is then an *omnivore* which spends all night searching and digging for small food items.

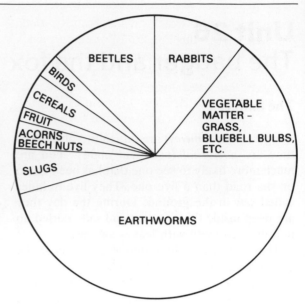

A badger's diet (typical annual intake; the actual amount will vary from year to year)

Task 26.1

From what you know of the badger so far, write down which of its sense organs are likely to be the best developed, and which least well developed. Give reasons for your answer.

Badgers do not hibernate; they remain active throughout the winter. When food is plentiful they build up reserves of fat. These help them to survive during periods of heavy snow and frost when they may be unable to find enough food.

Musk-glands

The badger belongs to the family Mustelidae; this includes stoats, weasels, polecats and skunks. All of these animals have glands under the tail which produce *musk* – a fluid with a very strong smell. Badgers' droppings smell strongly of musk and are used to mark out the boundary of their *territory*. This serves to warn badgers from other sets not to enter the area and begin feeding. Musk helps the badger to recognise members of its family. Musk is also used to mark the path on an outward journey when in strange territory so the animal can find its way back.

Breeding

For many years, naturalists were not able to discover the badger's gestation period. It seemed to vary between 2 months and over a year. It has since been discovered that, after mating, the fertilised egg remains inside the mother, but

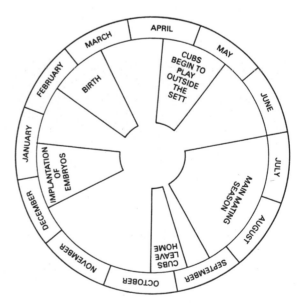

The badger's year

unattached to the womb and without a proper blood supply. In this condition the embryo does not develop (or develops only very slowly). In winter when the days are shortest the embryo becomes *implanted* inside the *uterus* (womb); then it begins to develop in the normal way. About 8 weeks later the young are born.

Although mating can take place at any time between February and October, the young are always born in February and March. This gives them the whole summer to grow and develop before leaving their parents in October. Badgers usually have two or three young at a time. These are born blind and helpless, but they develop quickly and are *weaned* (stop suckling) at about 12 weeks old.

Badger matings

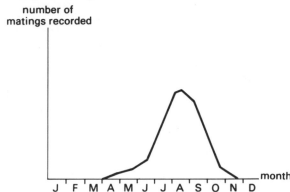

Badger births

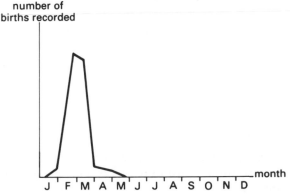

Badgers and people

The only enemy the badger has is people; in the past we have dug out badgers and killed them in the name of sport. Happily, this is now illegal. Badgers do eat a small amount of cereals and flatten a good deal more in the process, but modern machinery can harvest laid corn so no damage is done. They eat a little bark from trees, but not enough to damage the tree. To their credit, they eat a lot of rabbits, mice and rats – all of which are pest species. On balance the badger must be a beneficial animal to the farmer. It certainly gives much pleasure to the armchair naturalist and the more active naturalist.

Badgers will destroy rabbit fencing if it is laid across their track. The forester recognises the value of the badger and inserts a *badger gate* when fixing rabbit netting across a badger track. This consists of a swinging door which badgers can push through but rabbits cannot.

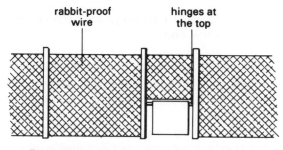

Front view

Side view   *A badger gate*

Practical 26.1

Using any materials you have available [card, netting, etc.], make a model of a badger gate. Remember it must open both ways as the badger usually returns along the same path.

There is concern about the numbers of badgers which get killed on the roads especially during October when the full-grown young leave their parents. New roads and motorways have special tunnels underneath when they are being built in areas frequented by badgers. The badgers soon learn to use these instead of crossing the road.

It is believed by some people (but not by others) that the badger infects cattle with a disease called *tuberculosis*. Some badgers in parts of Gloucestershire and the south-west counties of England have the disease; some cattle in these areas also have the disease. The Ministry of Agriculture has gassed many badgers in an attempt of wipe out the disease. This has led to fierce arguments between the ministry and nature lovers. As yet there is no convincing scientific proof that badgers either do or do not give the disease to cattle.

Model of a badger gate

Task 26.2

Read the following passage and answer the question at the end:

Badgers were surveyed in a large wood surrounded with farm land. Six sets were found, also five areas where there were numbers of dung pits.

Food consisting of peanuts, honey and coloured plastic beads was put just inside one of the entrances to each set. Different coloured beads were used for different sets. The dung pits were examined daily for plastic beads. The following records were made:

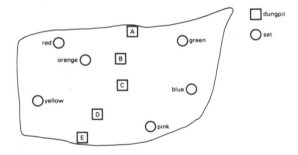

Dung pit: A	B	C	D	E
Day 1	No plastic beads found			
Day 2	Green	Red	Green	Green	Red
Day 3	Green	Red	Pink	Pink	Red
Day 4	Pink	Yellow	Pink	Pink	Yellow
Day 5	Pink	Orange	Blue	Blue	Orange
Day 6	Blue	Orange	Blue	Blue	Orange

The records were transferred to the map. Lines joining the sets to the dung pits where the beads were deposited were drawn.

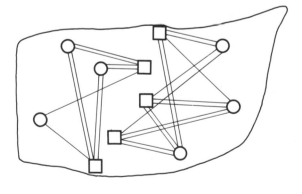

What can you learn about badgers from this investigation? Turn to page 148 and check your answer.

The fox

The fox has learned how to survive in cities as well as in the countryside and you are much more likely to see a fox than a badger. One reason for this is that they are not so strictly nocturnal as the badger; another is that they have a larger range.

Food

Foxes lie up in a hole (or underneath a city building) during the day and they hunt at night. They will take rabbits, rats, birds or almost any live prey they can catch. When they have young they drag the catch back to the den, leaving the unwanted parts outside in a very untidy manner. If a fox breaks into a chicken pen it will kill a lot of birds before making off with just one. In towns, urban foxes will raid dustbins, bird tables and pet food containers, as well as taking food which animal lovers put out for them. In addition to live animals, they take carrion and small animals like earthworms. These are freely available in cemeteries, parks, lawns and golf courses, as well as fields in the countryside.

Breeding

Vixens (♀) and *dog foxes* (♂) form pairs in January; they have an average of five *cubs* in

April. After the first few weeks of suckling and small amounts of dead food, the cubs are fed with injured animals which helps to teach them to kill. Later they go on hunting expeditions as a family group. When they are 6 months old, the cubs are fully grown (and trained). Then the whole family breaks up and they all lead solitary lives until the next mating season.

Task 26.2

1. Look at the 'badger year' diagram on page 144.
2. Make a similar one to show the life cycle of the fox.

Foxes and people

People have hunted foxes for many years, and still do. This *hunting* is more likely to conserve the fox than it is to reduce its numbers. Some people think of fox hunting as a sport and they are interested in keeping a good fox population so they have some animals to hunt.

In the main sheep-raising areas, foxes are shot by organised groups of farmers. This is because the fox takes newborn lambs. The author once saw this happen. A ewe had a lamb, licked it dry and allowed it to suckle. She then lay down and had a second lamb. Throughout this time a fox had been lying under a hawthorn bush nearby. As soon as the second lamb appeared the fox made a dart for the older lamb; this bleated loudly and ran. The ewe then rushed off to protect its lamb and the fox doubled back to the newest lamb, which it dragged away. The ewe was so busy protecting its first lamb that it didn't notice this happening.

The fatal disease *rabies* has spread across Europe from the East; there is a lot of scientific evidence that it is spread mainly by foxes. Great Britain has quarantine laws; these make sure that animals imported into this country are kept for several months in isolation (quarantine) to make sure that they are disease free. These laws have kept rabies (and other infectious diseases) out of this country. It is important that people obey these laws because if rabies spread to this country then a lot of foxes would have to be killed to stop it from spreading further.

Summary

1. Badgers are *nocturnal*.
2. Badgers live in holes called *sets*.
3. Badgers are *omnivores*; their chief food is earthworms.
4. Badgers do not hibernate, but they do build up fat reserves in autumn; these get used during the winter.
5. *Musk glands* under the tail produce a strong-smelling scent.
6. Badgers have a long mating season and a short birth season; this is possible as the embryo has a period of delayed *implantation*.
7. Badgers do a lot of good and very little harm.
8. *Badger gates* allow badgers through but not rabbits.
9. The Ministry of Agriculture believes that badgers give the disease *tuberculosis* to cattle. However, this has not been scientifically proved; many people disagree with it.
10. Foxes live in both town and country.
11. Foxes are carnivores.
12. A fox will take a new-born lamb.
13. Fox hunting conserves foxes.
14. On the European mainland it has been shown that foxes have been the main agent in spreading the disease rabies.
15. Quarantine laws in this country have so far prevented rabies from spreading from the mainland.

Questions

1. Write single sentences to answer the following questions:
   (a) Which animal family does the badger belong to?
   (b) How does the badger catch rabbits?
   (c) How do badgers stop wasps from stinging them?
   (d) At what age do young badgers leave their parents?
   (e) How does the badger mark out its territory?
   (f) What parts of a badger's food stay undigested?

2. (a) Using words and diagrams, describe a badger gate.
   (b) What is the purpose of a badger gate?
   (c) Describe one method of reducing badger deaths on the road.

3. (a) Describe the life cycle of the badger.
   (b) Describe the badger's diet. What four methods are used to obtain this information?

4. (a) List the items which a fox may eat.
   (b) Draw two columns and head one 'Town Fox' and the other 'Country Fox'. Transfer the items from your list into the correct columns; if an item will fit into either column write it down in each column.

5. Read the two statements:
   Statement one: Hunting conserves foxes.
   Statement two: Fox hunting is cruel.
   Write one of the following essays:
   Fox hunting should be banned. Fox hunting should not be banned.

Answer to Task 26.2

You should have learned the following:
1. A family of badgers can occupy more than one set.
2. The wood had two separate families of badgers.
3. Each family occupied (or visited) three sets.
4. Badgers use the same dung pits a number of times.
5. The dung pits *probably* mark the dividing line between the two groups' territories.

Unit 27
The rabbit

The rabbit (*Oryctolagus cuniculus*) was intro-duced to this country by the Normans. The rabbits were kept in a series of burrows called *warrens* to provide meat and fur for the Nor-mans. The rabbit provides meat to animals other than humans; foxes, badgers, stoats, hawks, owls and many other wild creatures feed upon this herbivore.

The rabbit food chain

Food

The main food of the rabbit is grass. This is very hard to chew and many herbivores and omni-vores (including people) are unable to digest it. Inside the rabbit is a large organ which contains lots of bacteria. The bacteria feed on the grass which the rabbit has eaten, breaking it down and producing millions more bacteria.

The partly digested grass together with the

bacteria form into a special oval-shaped drop-ping which is passed out through the anus. The rabbit then eats these droppings (called *copro-phagous pellets*) to obtain the protein it needs to balance its diet and remain healthy.

The rabbit uses its normal droppings to mark out its territory. Rabbits seldom wander more than 250 metres from home; they feed on plants growing nearby. This results in many plants getting overgrazed. Plants which the rabbit does not eat flourish in the absence of any competi-tion and many warrens become surrounded by nettles and thistles.

Rabbits do not only eat wild plants. They also eat crop plants like cereals and cabbages. In winter they will eat the bark of sapling trees. Loss of bark will kill a tree; tree guards similar to the one in the photograph on page 150 are fitted to protect trees from rabbits. Tree guards

are designed to protect the bottom $1\frac{1}{2}$ metres of a tree. Although a rabbit cannot usually reach this high it may be able to do so following heavy snow.

Breeding

Male rabbits are called *bucks*; female rabbits are called *does*. A doe will begin breeding when she

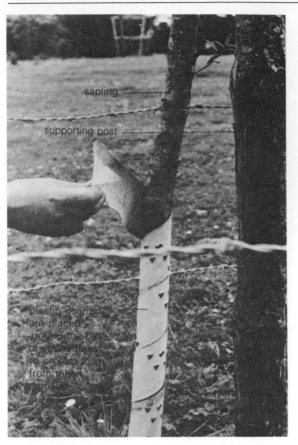

A tree guard

Myxomatosis

In 1953 a rabbit disease called *myxomatosis* was introduced into this country and almost all the wild rabbits died. One effect of this was to increase farmers' yields. However, the loss of rabbits also had bad effects. Many wild creatures had abundant food as dead and dying rabbits littered the countryside. This extra food was quickly followed by a famine as the rabbit disappeared. The numbers of several birds of prey began to fall. Creatures like the badger and the fox which have a more varied diet had to adapt to a diet without rabbit. It is likely that the loss of rabbit as a food source drove many foxes into the towns to scavenge dustbins and bird tables (see Unit 26).

A few rabbits survived and gradually the population increased. Myxomatosis returns every few years but rabbits are now more resistant than they were in 1953; so less die from this disease.

is only 3 months old. She may have as many as 30 young in one year. The *litter* size varies from three to eight at a time. A rabbit has to spend a lot of time away from its burrow eating, and only a few of those born will survive long enough to breed. The pregnant doe digs a burrow with only one entrance a short distance from the main warren. She makes a nest at the end of this burrow with dry grass and leaves. She lines it with hair plucked from her own underside. Each time she leaves the nest to feed, the doe covers the entrance to the hole with soil to protect her young whilst she is out feeding.

Control

The rabbit is still hunted for food; they are also shot and trapped for this purpose. One trapping method is to cover all the holes in a warren with nets and then to drive the rabbits out by letting a *ferret* (a small mammal similar to a stoat) into the burrows. This method is not easy, however; rabbits dig secret escape holes which are difficult to find.

Shooting and trapping rabbits for food has little effect upon their numbers. It is therefore not a very good way of controlling them. Control is carried out by experts who kill them in their burrows with cyanide – a poisonous gas.

If you live in an area where there are a lot of rabbits, the best method of protecting your garden crops is to have a rabbit-proof fence all around your garden.

Summary

1. The rabbit was introduced into this country by the Normans as a source of meat.
2. Most mammals are unable to digest grass; the rabbit can with the aid of *bacteria*.
3. Rabbits produce *coprophagous pellets*, which they eat.
4. Rabbits are herbivores.
5. Rabbits damage crops and young trees.
6. A *doe* may have 30 young in one year.
7. Rabbits are an important food source for some mammals and birds.
8. *Myxomatosis* is a disease of rabbits.
9. Hunting rabbits is not a very effective method of control.

Questions

1. Write single sentences to answer the following questions:
   (a) What is myxomatosis?
   (b) What do coprophagous pellets contain?
   (c) In what *two* ways do rabbits damage young trees?
   (d) Where docs a doe have her young?
   (e) What is a warren?
   (f) Why were rabbits introduced into this country?

2. Explain:
   (a) how a colony of rabbits can change the flora (plants) of the area around their warren;
   (b) how a rabbit can digest grass;
   (c) why myxomatosis is less important now than it was in 1953.

Glossary

Altitude The height above sea level.

Amphibian A cold-blooded vertebrate that changes from a water dwelling creature, breathing with gills, to a land dweller breathing air.

Anemometer Instrument that measures wind speed.

Annual A plant that completes one life-cycle in one growing season.

Anterior Head-end of animal.

Anther The part of a stamen that produces pollen.

Arable Cultivated land.

Arboreal Tree living.

Auricle Small extension of grass leaf blade.

Auxin Growth-regulating chemical produced in plants.

Awn Spike on the seeds of some grasses, e.g., barley.

Biennial A plant that requires two growing seasons to complete its life cycle.

Bird A warm-blooded vertebrate covered with feathers.

Brashing-up The removal of the lower branches of conifers.

Broilers Table chickens reared for the killing at an early age (10–14 weeks).

Brood chamber A box of frames in a beehive in which the queen lays eggs and young are reared.

Brooder Equipment for keeping baby chickens warm.

Broody hen Hen in condition to incubate eggs and rear chicks.

Browsing The eating of leaves, shoots etc., from the tops and side branches of trees and other large vegetation.

Buck Male rabbit and deer.

Bud A compact cluster of tiny leaves (or flowers) on a very small stem.

Bulb Roughly spherical plant structure having a reproductive and overwintering function, consists of layers of swollen leaves on a very short stem, e.g. onion.

Caecum Part of the gut of birds and some other animals.

Caesarean section Birth by cutting baby from mother's womb with scalpel or similar instrument.

Calcium carbonate A chemical substance from which egg shells are formed.

Calyx A ring of sepals on a flower or fruit.

Candling Looking into eggs by means of a bright light.

Canopy The area covered by the spread of a tree's branches.

Cardinal points The points of a compass N, S, E, and W.

Carnivore An animal that lives by eating other animals.

Caudal disc Area around the base of the tail.

Chalaza Twisted, thickened area in egg white which holds the yolk in position.

Clutch One sitting of eggs.

Coccidiosis Disease of young chicken, caused by a parasite – symptoms: blood in the droppings.

Compost 1. Decaying organic matter.
2. Material for filling seed boxes and plant pots.

Conifer Trees that bear cones (Yew and Juniper have berries, but are also conifers).

Contractile roots Strong roots on some corms and tap roots that contract and pull the plant deeper into the soil.

Coprophagous pellets Special droppings produced by rabbits at night which are immediately eaten.

Corm Underground swollen base of stem, stores food and produces new shoots from its buds.

Corolla The complete whorl of petals on a flower.

Cotyledon (a) The part of a seed in which food is stored.
(b) The first leaf produced by a seed.

Crop Storage organ in bird for undigested food.

Cross pollination Pollination of a flower with pollen from another flower.

Crown The uppermost parts of a tree.

Crown board Cover on top of a beehive to prevent bees entering the roof space.

Culm Flower stalk of grass.

Cutting Part of a plant which has been cut off to grow a new plant.

Deciduous Tree or shrub that bears no leaves in winter.

Dewlap Fold of loose skin hanging from the neck of cattle.

Diameter The distance across a circle, measured through the centre.

Dormant Alive but not growing or changing in any way.

Dorsal Back part of an animal.

Down Small fluffy feathers growing underneath the main feathers on ducks and geese.

Drake A male duck.

Drey The nest of a squirrel.

Duodenum The part of the intestines nearest to an animal's stomach.

Egg tooth Hard tip on bird's beak used to break out from the shell.

Embryo Baby plant or animal inside seed, egg or parent.

Enzyme A substance in animals which causes chemical changes to take place.

Ephemeral Short-lived.

Etiolated Plant grown with long internodes, due to insufficient light.

Faeces Undigested material passed from an animal's gut.

Fawn Young deer.

Ferret Half tame variety of pole-cat kept for driving rabbits from burrows.

Fibrous (a) Consisting of many fibres (e.g., peat).
(b) Fibrous root – slender root, often with slender branches.

Filament The part of a stamen that holds the anther.

Fish A cold-blooded vertebrate, covered in scales, confined to an aquatic environment.

Flower The sexually reproductive part of a plant.

Fruit The fertilised ovary of a flower containing seeds.

Function Purpose.

Fungicide Chemical substance that kills fungi.

Gander Male goose.

Geotropism The response to gravity by part of a plant.

Germination The breaking of dormancy by the embryo in a seed.

Gizzard The part of a bird's stomach that crushes the food.

Glume Protective layer around the flowers in a spikelet of grass.

Gosling Baby goose.

Gramineae Grass family.

Grazing The eating of grass, or other vegetation, from the ground.

Hay Sun-dried grass.

Herbaceous plant Soft stemmed plant – as distinct from trees and shrubs.

Herbicide Chemical substance that kills plants.

Herbivore An animal that lives by eating plants.

Hermaphrodite Individual plant or animal that has both male and female organs.

Hexagon Six-sided figure.

Hibernation An extended period of sleep during the winter months when food is not available.

Humidity The amount of water vapour in the air.

Hybrid Animal or plant produced by crossing two separate pure breeds.

Incubator Apparatus designed to keep eggs at constant temperature and humidity until the young hatch from them.

Inflorescence Flowering shoot containing many individual flowers or groups of flowers.

Ingest Eat.

Inorganic Substance that has never lived.

Insect An invertebrate with three body parts and six jointed legs attached to the middle part of the body.

Insecticide Chemical substance that kills insects.

Insectivore An animal that lives by eating insects and other invertebrates.

Internode The part of the stem between one node and the next.

Intestine The part of an animal's digestive system where food passes into the bloodstream.

Invertebrate An animal without a backbone.

John Innes compost Mixture of loam, coarse sand, peat, and fertiliser for growing plants in pots and boxes.

Kidney Animal organ that filters waste products from the blood.

Larva (larvae) The feeding stage between egg and pupa in the life cycle of many invertebrates. A larva usually looks very different to the adult form.

Lemma Part of the case that protects the sexual organs in a grass flower.

Ley A crop of grass or grass and clover mixture sown to last a limited number of years (1, 2, 3 or 4).

Ligule Small transparent vertical projection on a grass sheath where leaf-blade commences.

Liver Animal organ that stores food and produces bile.

Lodicule Very small part of a grass flower.

Macro-organisms Living things large enough to be seen with the naked eye.

Mammal A hairy vertebrate that suckles its young.

Mandible Mouth part of insect.

Medullary rays Food storage cells in wood.

Membrane A single layer of skin.

Micrometer screw gauge An instrument for accurately measuring the thickness of materials that are too thin to be measured with a ruler.

Micro-organisms Living things too small to be seen with the naked eye.

Micropyle Tiny hole in the testa of a seed through which fertilisation takes place.

Mineral salts Inorganic chemical substances.

Monoculture The growing of a single crop species over a large area.

Nectar Sweet, watery substance produced by flowers to attract insects.

Node The place on a stem where the leaf stalk is attached.

Nursery Area where young plants or trees are raised.

Omnivore An animal that has a mixed diet (part plant and part animal) e.g., man.

Organ Part of an animal or plant that performs a special function e.g., leaf or lungs.

Organic Substance that is living or has lived.

Ovary The part of a flower that develops into a fruit.

Ovule The part of a flower that develops into a seed.

Palea Part of the case that protects the sexual organs in the grass flower.

Pancreas Organ in animals that produces enzymes.

Perennation The survival of a plant from year to year.

Perennial A plant that flowers each year.

Petal One of the parts of a flower – often brightly coloured.

Petiole The leaf stalk.

Phototropism The response to light by part of a plant.

Plumule Small shoot inside, or just emerging from, a seed.

Pollen The male sex cells produced by a flower.

Pollination The transfer of pollen to the stigma of a flower.

Porter bee escape Metal applicance which allows bees to pass from one section of the hive to another, but not to return.

Posterior Tail end of animal.

Preen gland Oil-producing gland on a bird's back just anterior to the tail.

Prevailing wind The most frequent wind direction.

Propagation Increasing the numbers of plants from seeds or parent stock.

Propolis Sticky yellow substance collected by bees.

Proventriculus The enzyme producing part of a bird's stomach.

Pullet Female hen under 18 months old.

Pupa The non-feeding developmental stage between larva and adult of many invertebrates.

Queen excluder Sheet of metal punched with holes large enough for worker bees to pass through but too small to allow queens or drones to pass through.

Rabies Fatal disease of mammals.

Radicle Small root inside, or just emerging from a seed.

Reptile A cold-blooded, air-breathing vertebrate, covered in scales, which reproduces by laying eggs on land.

Rhizome Underground stem (usually horizontal).

Rodent Gnawing mammal with continually growing incisors.

Salmonella Micro-organism that causes food poisoning.

Sapling A young tree.

Seed A tiny dormant plant with a store of food surrounded by a protective coat.

Self-pollination Pollination of a flower with its own pollen.

Sepal The outermost, leaf-like structure of a flower.

Shrub A moderately sized woody perennial with many branches at its base.

Silage Partially decayed, stored grass.

Sink Wedge-shaped cut in a tree made as first part of the felling operation.

Smoker Beekeeper's appliance for puffing smoke.

Spikelet A cluster of individual flowers in a grass plant.

Stamen The male part of a flower.

Sterile Unable to breed.

Stigma The female part of a flower that receives pollen.

Stimulus A change in the external environment of plant or animal which produces a response in that plant or animal.

Stolon Horizontal stem that grows above the ground along the soil surface.

Stool The roots and lower stem of a perennial plant.

Straw Dead leaves and stems of cereal crops left after the grain has been removed.

Style The part of a flower between the stigma and the ovary.

Super Box of shallow frames in which bees store honey.

Syrup A strong solution of sugar in water for feeding to bees.

Tap-root A single main root that grows vertically downwards.

Tendril A slender outgrowth from a plant that curls around objects to give the plant support.

Terminal bud The bud on the tip of a branch or shoot.

Thigmotropism The response to touch by part of a plant.

Thinning The removal of some plants or trees to allow more space for the ones remaining to grow.

Tiller Bud of a grass plant.

Tropism A plant's response to a stimulus.

Tuber Part of a root or stem swollen with food stores.

Tuberculosis (TB) An animal disease, usually infecting the lungs.

Velvet Hair-covered skin, over developing antlers.

Vent The outlet in the back of a bird through which faeces and eggs pass.

Ventral Underneath part of animal.

Vertebrate An animal with a backbone.

Viable Having the ability to germinate (seeds).

Virus Extremely small form of life, that can live only in other living things.

Vixen Female fox.

Warfarin Poison used against rodents.

Weed A plant growing where it is not wanted.

Index